PETZOLD/RÖHRS

CONCRETE FOR HIGH TEMPERATURES

Concrete
for
High Temperatures

by

ARMIN PETZOLD and MANFRED RÖHRS

Technical University for Architecture and Civil Engineering, Weimar

Translated by

A. B. PHILLIPS
M. A., M. I. C. E.,
and
F. H. TURNER
B. Sc. (Eng.), M. I. C. E., A.M.I. Struct. E., D. I. C.

MACLAREN AND SONS
LONDON

85 334 033 1

CONCRETE FOR HIGH TEMPERATURES

BY ARMIN PETZOLD AND MANFRED RÖHRS

TRANSLATED FROM THE GERMAN

BETON FÜR HOHE TEMPERATUREN

2ND EDITION PUBLISHED 1967 BY

VEB VERLAG FÜR BAUWESEN BERLIN

VLN 152 · Dg.-Nr. 905/38/69 · DDR

THIS ENGLISH TRANSLATION

BY A. B. PHILLIPS AND F. H. TURNER

FIRST PUBLISHED 1970 BY

MACLAREN AND SONS LTD

7 GRAPE STREET

LONDON WC 2

Preface

The progressive development of civil and structural engineering leads to the continuous introduction of new materials and new construction methods. This phenomenon, which is especially marked in the case of the construction of industrial structures and blocks of flats, is now also becoming increasingly apparent in other branches of the industry, including the construction of industrial furnaces. To the pre-fired bricks which, until a few years ago, constituted almost the only building material used in this class of work, there has now been added a new material; this is concrete that is specifically suitable for high temperatures. The use of precast concrete units also makes it possible to adopt industrialised methods of construction.

Heat-resistant and fire-resistant concrete is treated in an unsatisfactory manner in those technical books which discuss refractory materials. This treatment is quite out of proportion to the importance which these materials have now attained and will continue to attain; in particular, it takes no account of the great variety of possible applications of concrete and the notable economic advantages which it possesses in comparison with some of the fired bricks which are in use. The authors therefore feel that they are right in offering a work devoted specially to concretes for use at high temperatures, where the whole subject is covered in a condensed manner in one book.

The development of concrete for high temperatures (or refractory concrete, as it is often known for the sake of simplicity) has extended during the last decade far beyond the use of normal industrial cements. In highly industrialised countries today concretes made with special cements or waterglass are now numbered among the common fire--resistant materials. New binding agents, such as magnesia binders, phosphoric acid, or phosphate, are becoming commoner in a wide range of applications.

The present stage of development in this field of high temperature concretes is reflected in the present work. The authors have endeavoured to take account of the speed with which progress is being made and to provide as complete a picture as possible of the situation as it exists today. Owing to limitations of space, however, it is not possible to go into full details in every case; a considerable amount of material had, therefore, to be abbreviated, leaving in a guide to the essential points. The greater part of the book has naturally been devoted to those heat-resistant and fire-resistant concretes which are made with cements, since these are still by far the most common materials in this field; for this reason questions of concrete mix proportions, concrete mixing and working into place are also dealt with solely by examples relating to cement-bound concretes. The treatment of similar questions for the newer types of concrete made with waterglass, magnesia, or phosphate binding agents, together with details of the practical applications of these materials, are subjects for further development.

The authors wish to express their thanks to a large number of individuals and firms for the many discussions, consultations and critical assessments which have assisted them and for the permission given to draw upon their work. Particular mention should be made of Dipl.-Ing. *E. Kuntsch*, of the VEB Silica and Chamotte Works, Rietschen, Doz. Dr.-Ing. *S. Röbert*, Doz. Dr. *G. Kriegel*, both of Weimar, Dipl.-Ing. *H.-J. Golde* and other members of the staff of the VEB Development Section for Ordinary Ceramics, of Meissen, and also of the VEB Chamotte Works, of Brandis.

Finally, the authors wish to thank the VEB Verlag für Bauwesen (the publishers of the original German edition) and in particular Herr *E. Peukert*, for their most valuable help and support in the writing and editing of this work.

Armin Petzold   *Manfred Röhrs*

Preface to the second edition

The speed with which the copies of the first edition of this book were taken up, and the excellent reception accorded to it both in the German Democratic Republic and abroad, are evidence of the great interest which is now shown in refractory concrete. The first edition was consequently sold out shortly after its appearance. In order to do justice to the continuing demand for the work, the Authors and Publisher decided that they should soon follow it with a second, improved and enlarged edition. The sections dealing with the composition of concretes and the manufacture of lightweight fireproof concretes have been revised and a number of new concrete mixes has been introduced. The chapter on the properties of concretes is also extended to include a section on the behaviour of heat-resistant reinforced concrete. Finally, the chapter illustrating the practical uses of concretes in plants which operate at elevated temperatures is enlarged by the addition of a number of new examples.

The Authors

Contents

1. Introduction

Until a few years ago the fire-resistant linings of furnaces and similar industrial plants operating at elevated temperatures were constructed almost exclusively of precast and prefired bricks of relatively small dimensions and all the various aspects of design, construction and operation of such linings were considered in this context. The principal exceptions were glass trough bricks of large dimensions and in-situ products intended for use as mortars, rammed concretes, sprayed concrete and renderings for repairs and similar surface treatment. Since the last-named materials, according to *K. Konopicky* [1], constitute only 12 to 15% of all materials of this class used in fire-resistant work, the total proportion used in all large-scale fire-resistant construction, including, for example, hearths, furnaces, etc., must be appreciably smaller. In the last 10 years, however, the increased use of concrete has resulted in the proportion of unfired materials being modified, so that in the U.S.A. today, for example, the production of fire-resistant concrete is estimated at approximately 9% of all fire-resistant materials.

Although the firing process of fire-resistant materials is ensured by the operating temperatures of the plant itself, prefiring of the major part of the raw materials and of the bricks themselves for a long time proved to be almost indispensable, if the required properties of the final product, such as strength, density, slag resistance, and so on, were to be achieved and a large number of difficulties relating to thermal behaviour, such as growth, shrinkage, and so on, were to be avoided.

The trend towards the elimination, in the last stage, of the heat-treatment process, thereby effecting economies in production costs, has been continued now for several decades. The principal means of achieving this has been first by developing unfired bricks or precast components which possess the necessary load-bearing capacity, while at the same time having a sufficiently low degree of shrinkage, and secondly by the increased use of mixes and mortars which are capable of being worked on site into monolithic structures. The last-named materials must also possess both high strength and a negligible tendency towards shrinkage. Typical examples of these developments are

the unfired basic bricks ("chemically bound" products of the magnesia-
-chrome ore range and, to a lesser extent, also tar-dolomite bricks)
and the various refractory concretes. While, however, the former
group of materials usually requires the application of high pressure
for the formation of components and these components are limited
to standard sizes, the use of concrete permits the production of precast
units of large size or the in-situ construction of monolithic walls and
similar structures which achieve high structural strength at an early
age.

The development and application of the technology of high temperature concretes follows a similar course to that of concrete used in
conventional civil engineering; in the same way as progress has been
made from traditional brick construction towards the use of blocks
of large dimensions in the building of large areas of walling capable
of standardisation, using either concrete or ceramic-concrete compounds (structural earthenware or pottery) as the material, so have
normal standard bricks been replaced to an increasing extent by
precast units or the in-situ concreting of complete walls, bases and
hearths, when fire-resistant linings are required for industrial furnaces.
This development is natural and logical and will spread to every area
of work in which it offers advantages over conventional lining construction methods with regard to rationalisation, productivity, increased output and economy.

1.1. Terminology and Classification

A lack of uniformity exists in the descriptions applied to concrete-type materials intended for use at medium and high temperatures. Similarly, there is no standard which defines the materials to be included within the scope of this classification. It is therefore essential to begin with some remarks on the descriptions which are used for concretes which are resistant to high temperatures.

1.1.1. Problems of Nomenclature

The terms "feuerfester Beton" (fire-resistant concrete) and "Feuerbeton" (refractory concrete) are in general use today when referring to thermal properties (*K. Konopicky* [1], *F. Harders* and *S. Kienov* [2], *G. Rabe* [3], the standard of the German Democratic Republic TGL 9356 [4], TGL 99-30 [5]). In the Soviet Union the term "hitzebeständiger Beton" (heat-resistant concrete) is used as an overall description (*K. D. Nekrasov* [6], [7], [8], [9]); in the Soviet Standard GOST 4385-48 a distinction is made between "hochfeuerfesten Betonen" (high-temperature resistant concretes) (resistant to temperatures $> 1770\,^{\circ}\mathrm{C}$), "feuerfesten Betonen" (fire-resistant concretes) (resistance 1580 to $1770\,^{\circ}\mathrm{C}$), and "hitzebeständigen Betonen" (heat-resistant concretes) (resistance $< 1580\,^{\circ}\mathrm{C}$) [8][1]). In English and American literature similar materials are described as "refractory cements", "refractory concrete", "fire cements", or "refractory castables" and in France as "bétons réfractaires".

With regard to composition, concretes for use at high temperatures may contain as aggregates all materials which are well known for their fire-resistant properties and are normally used for industrial purposes. In addition to normal and special cements waterglass and other cementitious substances may be used as binders. In this context no clear distinction is drawn between "Beton für hohe Temperaturen" (concretes for high temperatures) and "hitzebeständiger bzw. feuerfester Beton" (heat-resistant or fire-resistant concrete).

In the more important contemporary literature on the subject, such as books, periodicals and patent documents, it is often impossible to find any clear lines of demarcation between descriptions; the spectrum of the materials provided by the individual authors ranges from typical concrete (see below), through granular and fine-grained mixes using cold binding agents other than cement, to actual tamped concretes and mortars. Some authors, however, have seriously attempted to establish unambiguous terminology for high-temperature concrete (*P. P. Budnikov* [10], *F. Zapp* and *F. Dramont* [11], *H. Mitusch* [12], *A. Braniski* [13]).

Since a unique description based upon the thermal characteristics of the material under discussion evidently does not exist and since, moreover, only the description "fire-resistant" (feuerfest) is defined and there are no rules of

---

[1]) In the U.S.A., ASTM C 213—47 T classifies concretes as follows:
concretes for high operating temperatures (up to 1480 °C),
concretes for medium operating temperatures (up to 1315 °C).

nomenclature for grades of thermal resistance below this level it is naturally difficult to introduce a general term for all concretes designed for temperatures above 300 °C. "Concretes for High Temperatures" is in essence a correct term, perhaps also the sub-division into heat-resistant and fire-resistant concrete. In addition to and in parallel with these descriptive terms, the term "refractory concrete" (Feuerbeton)[1] is used for the whole range of concretes because, although not grammatically correct in the strictest sense, it is justified by its simplicity and by the frequency of its use in the literature referred to above.

1.1.2. Definitions

The word "concrete" is generally understood to mean a mixture composed of cement, aggregates of many kinds and particle-sizes, and water, which hardens at normal temperatures [14], [15].

Normal, or heavy, concrete has a density of 1,9 to 2,5 gm/cm^3 (120 to 155 lb/ft^3), while that of light-weight concrete is between 0,3 and 1,6 gm/cm^3 (20 and 100 lb/ft^3). All materials not containing cement can be satisfactorily described as "concrete-type" materials.

The term "cement" describes finely-ground hydraulic binding agents, consisting principally of combinations of CaO with SiO_2, Al_2O_3 and Fe_2O_3, having certain characteristics laid down in Standards and reaching a minimum standard compressive strength after 28 days of hardening, of 225 kp/cm^2 (3200 psi); the raw mass or the major part of it must have been heated at least to the stage of sintering. The most important cements in this connection are portland cement, blast-furnace cement and the various aluminous cements; one of their most important distinguishing characteristics is hydraulic hardening.

Hydraulic binding agents are artificially produced mineral substances of very fine grain size (dust-like) which, when mixed with aggregates and water, form a mixture, the wet concrete; after a certain period — usually a few hours — this sets by the combination of the water and then continues to harden into a stone-like material, insoluble in water and resistant to the attack of water (*H. Kühl* [16]). The hydraulic hardening continues either in air or under water. The description "hydraulic" therefore refers to the characteristic of a binding agent which combines with water, hardens under water and remains water--resistant. The most important hydraulic binding agents and those with the greatest technical value are the cements.

In the terminology used in the refractory industries a material is described as fire-resistant only if its cone fusion test point (fire-resistance) is at least 1500 °C. This fact leads to the division of concrete-type materials also into fire-resistant concretes for which the test point > 1500 °C and into those which do not reach this value; the latter may conveniently be described as heat--resistant, rather than fire-resistant, concretes[2]. This follows in principle the divisions used in GOST 4385-48, and does not recognise a special sub-division for high-temperature resistant concrete (cone fusion test point > 1790 °C).

---

[1]) Translators' Note: This sentence relates, of course, to the German word "Feuerbeton".
[2]) In this connection, *R. Barta* [17] refers to "semi-fire-resistant" concrete.

The classification "fire-resistant concrete" is therefore taken to include material composed of cement and fire-resistant aggregates, whose cone fusion test point exceeds 1500 °C.

Heat-resistant concrete is material composed of cement and aggregates, which can be employed for use at temperatures higher than those normally encountered in civil engineering but whose cone fusion test point lies below 1500 °C. A lower temperature limit is not given.

Some engineering standards formerly in use defined substances composed of fire-resistant concrete granules which ". . . are hydraulically bonded at normal temperatures, gain mechanical strength and, when fired, pass from the hydraulically bonded to the ceramically bonded state without appreciable loss of strength", which, it will be noted, is possible only with smelted aluminous cement. Precise definitions for a number of such materials are given in TGL 9235 and TGL 99-30. According to these Standards concrete mixes for refractory-plants are air-dried mixes composed of fire-resistant raw materials of grain-size 0 to 30 mm (0 to 1,25 in) and of cement.

While the use of cement is specifically referred to here, the terminology used in the Soviet Union is not so exact. For example, *K. D. Nekrasov* [6] states ". . . heat-resistant concrete is the description used for a particular type of concrete which retains its physical and mechanical properties within certain limits even after long periods at high temperatures". According to the "Instructions for the manufacture and use of heat-resistant concretes" [8] the binding agents used in such concretes may be hydraulic binders (usually cements) or air binders (Periclase cement with Sorel's cement, waterglass); in the Soviet Union heat--resistant concrete with a waterglass basis plays quite an important role, as can be seen from a number of publications.

In the Western countries importance is attached to the use of hydraulic binding agents in high-temperature concretes, as can be seen from the definitions in the Vocabulary of Refractory Terms (*E. Stephan* [18]). This states that "castable refractory" means a substance which hardens hydraulically after mixing with water.

In accordance with the above definitions this work will deal chiefly with heat-resistant and fire-resistant concretes having a cement basis. The subject of concrete-type substances having a basis of waterglass, phosphate or magnesia (Periclase) binders will be touched upon only briefly and for the purpose of general information.

1.1.3. Classification Systems for High-temperature Concretes

High-temperature concretes may be classified according to any one of a number of different systems such as by operating temperature or fire-resistance, by the type of binding agent, or by the type of fire-resistant aggregate employed.

Classification according to fire-resistance, as mentioned above, gives the following groups of concretes:

(a) heat-resistant concrete   (cone fusion test point (cftp) < 1500 °C; operating temperature 200···1100 °C.)

(b) fire-resistant concrete

(cftp lying between 1500 and 1790 °C; operating temperature 1100···1300 °C)

(c) high-temperature-resistant concrete

(cftp > 1790 °C; operating temperature > 1300 °C).

Classification according to the type of binding agent gives the following sub-divisions:

(a) refractory concrete made with pozzolanic cement or slag cement (blast--furnace cement or iron portland cement);

(b) refractory concrete made with portland cement;

(c) refractory concrete made with aluminous cement:
   α) with normal aluminous cement,
   β) with aluminium oxide-rich aluminous cement (high alumina cement),
   γ) with barium aluminous cement;

(d) refractory concrete with waterglass binding agent;

(e) refractory concrete with a magnesia binding agent;

(f) refractory concrete with other types of chemical binding agents.

In a classification according to aggregate type the following system may be used:

(a) refractory concrete with aggregates which are not fire-resistant (broken brick, slag, tuff, etc.);

(b) refractory concrete with chamotte (burned fire-clay);

(c) refractory concrete with high alumina aggregates;

(d) refractory concrete with corundum;

(e) refractory concrete with silica;

(f) refractory concrete with magnesia;

(g) refractory concrete with chrome ore and chrome ore/magnesia;

(h) refractory concrete with silicon carbide.

While the fire-resistant properties of concrete can certainly be considerably influenced by the aggregate, a limit to their use is set in the last resort by the most easily fused constituent, in this case the binding agent. The system of classification according to the binding agent, which is widely used, will therefore be employed here.

1.2.   Historical Developments

According to *K. D. Nekrasov* [6] the first reference to the possibility of using concrete at high temperatures dates back to the work of the Russian structural engineering expert *N. A. Zitkevič*, at the beginning of this century. This writer had already realised that chamotte and other pre-fired materials were preferable to sand as an ingredient of concrete for this purpose, although his recommendations were not put to practical use at the time.

No serious attention was given to this new type of refractory material until 20 years later, when *C. Platzmann* patented the manufacture of refractory concrete based upon portland cement with the addition of chamotte and either trass or reactive silica [19]. The use of aluminous cement ("ciment fondu") goes back to the patent taken out by *P. Kestner* in 1925/26 [20].

The beginning of the 1930s saw a considerable increase in research and development in a number of countries. Among the contributions to the literature on the subject at this time the following should be mentioned: An anonymous author [21], *H. T. Coss* and *N. J. Cent* [22], *A. Custodis* [23], *A. Braniski* [24], *W. Czernin* [25], *V. A. Kind* and *S. D. Okorokov* [26], *I. A. Aleksandrov* [27], *S. V. Glebov* and *E. A. German* [28], *I. E. Gurvič* [29], *I. L. Movševič* [30], *A. I. Rojzen* [31], *G. M. Ruščuk* [32], *P. P. Budnikov* and *D. S. Il'in* [33], *A. V. Hussey* [34], *M. Lepingle* [35], *A. Möser* [36], and *R. T. Giles* [37], [38]. It was not until after the second world war, however, that the possibilities offered by heat-resistant and fire-resistant concrete really began to be explored. The large number of publications throughout the world is evidence of this and indicates the effort being made both in the realm of pure research and also in the development and application of the use of refractory concrete. Although not wishing to discuss these numerous works separately, the Authors feel they must mention the pioneer work of *K. D. Nekrasov* which has been responsible for the technical progress made in the U.S.S.R. in this field. His name is firmly linked with the subject of heat-resistant and fire-resistant concrete. Among his numerous publications testifying to this fact, the following papers and books may be particularly mentioned: [6], [39], [40], [41], [42]. *H. Lehmann* and *H. Mitusch* [43] have become well-known in Germany through their monograph on aluminous-cement-refractory concrete.

1.3.   Peculiarities and Implications of Refractory Concrete

The peculiarities of refractory concrete can be appreciated and explained either by comparison with normal concrete technology or by comparison with normal refractory technology.

To the concrete technologist the principal peculiarity consists in the use of special fire-resistant aggregates having definite characteristics, often certain very fine aggregates such as clay, and in the employment of aluminous cement, which is seldom used in concrete for other purposes, or of other even less common binding agents, such as waterglass and phosphate. The deviations from normal concrete technology are, however, slight and are hardly of importance. From this standpoint, therefore, refractory concrete can hardly be said to represent anything new. Whereas, however, definite values of structural strength are laid down for normal concrete, this is usually of secondary importance for refractory concrete, since the thermal strains which the concrete undergoes in service fundamentally alter its structure. The same is true of other characteristics to which importance is often attached in normal concrete.

From the point of view of the engineer dealing with refractory materials, refractory concrete presents definite peculiarities in both its manufacture and its use. Whereas the technology of refractory materials can normally be thought of in terms of, first, preparation, followed by formation to the correct shape in the plastic or half-dried state (principally by pressing) and then by long periods of drying and firing at high temperature, the technology of precasting refractory concrete comprises preparation of the mix, then working into shape by pouring and vibrating, hardening by an autonomous process (with retention of moisture) and finally a short period of drying. The necessary strengths are achieved without the process of burning and it is possible to fabricate large--sized units of varied and complicated shapes without danger of deformation and cracking. Refractory concrete for use in pyrotechnical plant may be either built up or cast in-situ as a furnace lining by normal concreting procedures. The advantages of monolithic walls and other structures in furnaces from many points of view are indisputable. One particular advantage is the immediate setting of cast-in-place units by a cold-setting process. In this respect refractory concrete differs fundamentally from normal fire-resistant mortars, pressed units and lining repairs. However, it must not be forgotten that linings of refractory concrete have strength characteristics which do not remain constant throughout all ranges of temperature, but invariably when first heated pass through the critical region of change from a hydraulic to a ceramic binding. If this region of change has not been exceeded under operating conditions, the strength of the statically-loaded structural elements has not been properly demonstrated. This phenomenon does not show itself with normal fired refractory bricks, except in the case of chemically bonded unburnt brick, for which it is essential that the critical region should be passed through and also that a minimum temperature should be maintained in use.

From what has been said the importance of refractory concretes in pyrotechnical plant can be clearly seen, both from a technical and an economic point of view. Even though the use of this material may occasionally be limited by the magnitude of the temperature — normal industrial refractory concretes do not yet withstand the effects of very high temperatures — the existence of certain concretes produced on an appreciable scale with special binding agents (such as SECAR 250) which are capable of being used even at very high temperatures, together with the rapid development in the Soviet Union of new concretes capable of withstanding very high temperatures, suggest that in the near future such limitations will no longer apply.

The importance of refractory concrete and the question of the economics of its manufacture and use are dealt with in Chapter 7.

2. The Basic Materials of Refractory Concrete

The main constituent materials of refractory concrete are the binding agent and the aggregate, the latter usually being fire-resistant. Very fine aggregates of various origins may also be necessary for special mixes, and porous materials for lightweight concretes. Water is the other main constituent and various additives may also be used at times.

In accordance with the definition given in Section 1.1.2, cement is the only binding agent used in the material described as "concrete". In this chapter the emphasis will be in the first place on the cements, since in the German Democratic Republic and in many other countries today the technology of refractory concrete is still based for the most part upon the use of hydraulic binding agents. The use of waterglass and the development of heat-resistant waterglass concretes for many purposes is, however, continually increasing, so that in certain fields it can be expected that this material may equal, or even predominate over, concrete made from cement in the foreseeable future. For this reason the use of waterglass as a binding agent will also be discussed; a summary will also be given of the new developments of concrete-type materials using other more or less common binding agents such as phosphoric acid and phosphate, magnesia binders, barium cement, and so on. All aggregates and additives will be discussed which are so far known to have been used in the making of heat-resistant and fire-resistant concrete.

In this chapter the constituents will be looked at solely in relationship to their chemical, physical and thermal properties. Special questions of granulometry, important in the manufacture of concrete, will be left to Chapter 3.

2.1.   Cements

Normal industrial cements are mineral products which have been very finely ground after sintering or smelting. They usually consist of crystalline or vitreous substances, the major components of which are CaO, SiO_2, Al_2O_3 and Fe_2O_3. They can be divided into silicate and aluminate cements. Typical examples of the first group are ordinary portland cement and the slag cements (iron portland cement, blast-furnace cement and supersulphated cement), and of the second group, the aluminous cements.

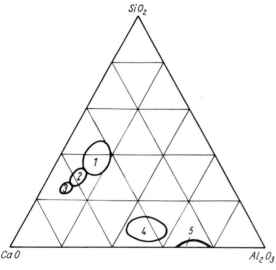

Fig. 1. *The position of the cements in the* CaO—SiO_2—Al_2O_3 *system*
1 Blast-furnace slag; 2 Slag cements; 3 Portland cement; 4 Aluminous cement; 5 High alumina cement

The constituents of overriding importance in the mineral structure of cements are CaO, SiO_2, and Al_2O_3; the part played by Fe_2O_3 (and by MgO and certain other secondary constituents) will be discussed for separate types of cement. The more common cements therefore may be described in terms of the three-constituent system $CaO-SiO_2-Al_2O_3$, as shown in the Rankine Diagram in Fig. 1.[1]

From the types of cement normally produced, the only ones used in the making of refractory concretes are portland cement, iron portland cement, blast-furnace cement, and normal (industrial) and special aluminous cement. At present other cements are of only minor importance or of scientific interest. The types just referred to are discussed below, with particular reference to their most important characteristics and their influence upon the manufacture of refractory concrete.

2.1.1.   Portland Cement

Portland cement is the prototype of a high-quality hydraulic binding agent, produced in mass-production processes. In the German Democratic Republic

---

[1] Supersulphated cement is not shown, as this material contains larger quantities of SO_3; it would lie at about the lower limit of the slag region.

the technical requirements for its supply are given in the Standard TGL 9271 [44] and the standard tests listed in TGL 10573 [45].

TGL 9271 defines portland cement as an hydraulic binding agent, which contains, after thorough, fine grinding portland cement clinker and gypsum or anhydrite stone; the sulphate content is limited to the amount permitted in the cement. Portland cement clinker consists of high-basic combinations of calcium oxide with silicon dioxide, aluminium oxide and iron (III)-oxide together with traces of magnesium oxide.

2.1.1.1. Chemical and mineralogical composition

The main constituents of portland cement are the minerals of portland cement clinker: for the sake of simplicity portland cement will be taken to comprise only these minerals, though in practice the addition of a slag component is permissible.

In the chemical composition of clinker the main constituents lie generally in the following ranges:

CaO	$62 \cdots 67\%$	MgO	$0,5 \cdots 5\%$
SiO_2	$18 \cdots 25\%$	Alkali	$0,5 \cdots 1,5\%$
Al_2O_3	$4 \cdots 8\%$	SO_3	$0,5 \cdots 2\%$
Fe_2O_3	$2 \cdots 5\%$		

MgO does not normally have any influence on the stability in the hydraulic phases. It does however have a very marked effect on the temperatures of sintering and smelting of the combined materials; the same is true of Fe_2O_3. This fact is of considerable importance in the fire-resisting properties of portland cement, as will be shown later.

For a full description of phase-stability portland cement should be considered in terms of the four constituents $CaO - SiO_2 - Al_2O_3 - Fe_2O_3$, or for still greater accuracy, in terms of five; namely $CaO - SiO_2 - Al_2O_3 - Fe_2O_3 - MgO$. For the sake of simplicity, however, the phase relationships are usually set out in the simpler three-constituent system: $CaO - SiO_2 - Al_2O_3$. From the arrangement of the portland cement section in the "calcium-corner" of the Rankine diagram repeated in Fig. 2, the theoretical phase equilibrium in the $C_3S - C_3A - C_2S$ triangle can be seen, for which the lowest temperature of crystallisation from the residual smelted mixture can be seen at the intersection point as 1455 °C. The corresponding relationships in the four-constituent system give a temperature of 1338 °C for the eutectic of C_3S, C_2S, C_3A and C_4AF.[1] If appreciable instability is present the result will be the occurrence of calcium-deficient aluminates ($C_{12}A_7$), free lime (CaO) and vitreous phases of varying composition.

The most important mineral phases of portland cement clinker are as follows:

Tricalcium silicate	$3CaO \cdot SiO_2$	(C_3S),
Dicalcium silicate	$2CaO \cdot SiO_2$	(C_2S),
Tricalcium aluminate	$3CaO \cdot Al_2O_3$	(C_3A),
Tetracalcium aluminate ferrite	$4CaO \cdot Al_2O_3 \cdot Fe_2O_3$	(C_4AF).

---

[1] C_4AF is in reality a mixed crystal of the hypothetical mixture range C_2F—C_2A, which begins at C_2F and ends near C_6A_2F.

Small quantities of the following may also be present, depending upon the nature of the raw materials and the conditions of clinkering:

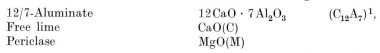

12/7-Aluminate	$12\,CaO \cdot 7\,Al_2O_3$	$(C_{12}A_7)^1$,
Free lime	$CaO(C)$	
Periclase	$MgO(M)$	

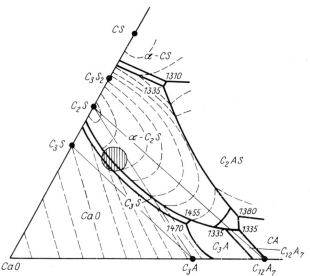

Fig. 2. *The position of port-land cement in the $CaO\!-\!SiO_2\!-\!Al_2O_3$ system*

C_3S occurs as a very important clinker mineral in solid solution with small percentages of C_3A and MgO and is known in this form as "Alite". C_2S, occurring predominantly in the β-modification, is known as "Belite". In normal portland cement the two silicates together add up to about 75 to 80% of the total mineral composition. C_3A (about 10%) and C_4AF ("Celite" or "Brownmillerite", about 5 to 8% of the mass of clinker) are often present with small quantities of other minerals in vitreous form and occur intermediately between the silicate crystals.

Portland cement can contain up to 0,1% by mass of chloride as a setting (accelerating) agent, in addition to gypsum; a higher content of chloride should be viewed with suspicion; it may have an important effect upon the fire-resistance of the cement. Constituents such as chloride, sulphate and magnesia have a softening effect upon cement which should not be underestimated.

2.1.1.2. Hardening of portland cement

Although the gross composition of portland cement is obviously not affected by hydration and hardening, there are wide variations in its mineral state which arise from the hydration process and, even more, from the temperature to which it is subjected. It is therefore useful to understand the main features of the hydration process.

---

[1]) $C_{12}A_7$ is often described in Soviet and American writings as C_5A_3.

The hydration and hardening of portland cement are distinguished by two fundamental processes associated with the principal component, calcium silicate: the formation of calcium silicate hydrate, of low lime content compared with the lime-rich clinker minerals, and the resultant occurrence of free lime in the form of hydrate of lime, $Ca(OH)_2$. The reaction with water causes the silicates to change to a gel-like phase, subsequently becoming crystalline and having the property of retaining water. The crystalline phases are usually of the tobermorite type, exhibiting a variable composition, since with the formation of the so-called "tobermorite-like phase", interstratification of $Ca(OH)_2$ between the tobermorite laminations — $Ca_5[Si_6O_{16}(OH)_2] \cdot 4H_2O = C_5S_6H_5$ — is possible. The surplus lime derived from the alite is deposited in the form of $Ca(OH)_2$ layers between the crystals of tobermorite.

Hydrates of varying composition arise from the hydration of the aluminates; examples are C_2AH_7, or, upon taking up more lime, C_4AH_{13}, and the most important of all, which is the most stable phase of aluminate hydrate, C_3AH_6. The formation of free combinations does not occur, however, in portland cement of normal composition, because complex combinations of $3CaO \cdot Al_2O_3 \cdot CaSO_4$ $\cdot 12H_2O(C_3ACsH_{12})$ and $3CaO \cdot Al_2O_3 \cdot 3CaSO_4 \cdot 31H_2O(C_3ACs_3H_{31}$, ettringite) arise through the reaction with gypsum in the process of the control of setting. These combinations are known as calcium aluminate sulphate hydrate, the last mentioned being the more stable.

The ferrite phase (principally of the combination C_4AF) reacts with water to form C_3AH_6 and calcium ferrite hydrates, which probably deteriorate to $Fe_2O_3 \cdot nH_2O$ and $Ca(OH)_2$.

C_3A and C_3S hydrate the most rapidly, C_4AF more slowly and C_2S slowest of all. The reaction is accelerated if the temperature is raised (steam treatment, autoclave hardening), but the effect upon the properties and particularly the strength depends upon the presence of aggregates, and also on their nature.

Hardened portland cement forms a stone-like material with a microcrystalline, partly gel-like, structure, in which both newly hydrated portions and partially hydrated particles of clinker are present. The proportion of water entrained in the newly hydrated parts is generally about 15 to 20%. This and the fact that for technical reasons the water/cement ratio (W/C) is always greater than the theoretical value of 0.25 (usually 0.5 to 0.6), leads to the consumption of water by evaporation and continuing hydration during the hardening process, resulting in the formation in the cement structure of hollows, large pores, and particularly capillaries and gel-pores. The structure of the hardened cement is very much dependent upon the hydration conditions (W/C ratio, temperature and aggregates). The porosity and permeability of the concrete and the quantity of entrained water are of considerable importance in this connection for refractory concrete, because the structure goes through a critical stage while the water is being driven off at first heating.

2.1.1.3. Technical properties of the hardened cement mass

The properties of portland cement which are important in relation to its use in fireproof concrete are naturally of the greatest interest when related to the concrete itself; some of the most typical properties of the pure cement will, however, be selected and discussed here in this connection.

The fire-resistance defined by the cone fusion test point is a direct function of the chemical composition and the related phase relationships in the equilibrium diagram. The extent to which the temperature of the first phase of melting of portland-cement-type combinations varies with the number of components in the combination can be seen from the following table, due to *R. H. Bogue* [45a]:

System	Commencement of melting, °C
$CaO - SiO_2$	2050
$CaO - SiO_2 - Al_2O_3$	1455
$CaO - SiO_2 - Al_2O_3 - Fe_2O_3$	1340
$CaO - SiO_2 - Al_2O_3 - Fe_2O_3 - MgO$	1300
$CaO - SiO_2 - Al_2O_3 - Fe_2O_3 - MgO - Na_2O$	1280

The cone fusion test points can be expected to lie approximately in these ranges of temperature and it therefore follows that normal portland-type cements barely reach the level defined as fire-resistant (test point 1500 °C). It is therefore evident that for portland cement to achieve maximum fire-resistance, it should be free from all secondary constituents, in particular MgO and Fe_2O_3.

This conclusion is confirmed by measurements of the cone fusion test point. Whereas ordinary portland cement has a cone fusion test point of around 1350 °C, the value for the special Dyckerhoff Weisz portland cement — class 225 — is given as 1500 °C [11].

According to Soviet sources the onset of softening under pressure lies in the range 970 to 1130 °C for pure cement and of unrestrained (unlimited) softening between 1350 and 1480 °C.

The pore structure and capillary structure of pure cement has already been mentioned in Section 2.1.1.2. For pure cement, assuming perfect compaction, capillary pores of diameter 10^{-4} cm and gel pores of only 20 to 40 Å in diameter are to be found. Nevertheless, the porosity which actually occurs may not be insignificant, because complete consolidation and filling of all voids can never be achieved for practical reasons. An overall porosity in the hardened cement structure of about 15% should therefore be expected in practice.

2.1.1.4. Alteration due to heat

If a portland cement structure is subjected to heat after it has set, it loses its adsorbed and entrained water in certain ranges of temperature and, after passing through an intermediate water-free state, finally reaches a stage where new phases are built up; the mineralogical composition of these is determined by the temperature. In this connection it is first of all important to obtain an understanding of the ranges of temperature in which dewatering and dehydration occur and of the process by which the water is lost. Strain-shrinkage behaviour and strength characteristics are dependent upon these factors.

The simplest methods of following the chemical changes resulting from heat are those utilising thermogravimetry and differential thermoanalysis (TGA and

DTA respectively). *Nekrasov* [6] in particular, and also *Lefol* [46], have published thermogravimetric measurements on neat hydrated clinker minerals. Important and more comprehensive investigations have been carried out by means of DTA. The results obtained by *A. Petzold* and *I. Göhlert* [47] are given in Fig. 3; they agree in their essentials with those of many other authors.

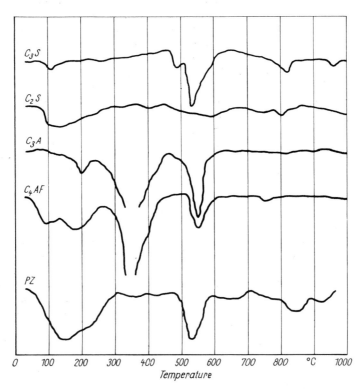

Fig. 3. *Graph showing the differential thermoanalysis of hydrated clinker minerals and portland cement (PZ)*

The endothermic effects indicated by DTA in the ranges 100 to 200 °C and 500 to 600 °C, which are attributable to the driving out of gel water and interlaminar water and the dehydration of $Ca(OH)_2$, are characteristic of C_3S. The third endopeak at about 800 °C can be ascribed to the dewatering of calcium silicate hydrate, while the fourth indicates the deacidification of $CaCO_3$ which has been produced. The significant exothermic effect between 800 and 900 °C is the result of the formation of wollastonite which takes place in this range of temperature.

The thermogram of hydrated β-C_2S is, by contrast, not at all typical. The absence here of the $Ca(OH)_2$-effect is striking.

The endothermic effect for hydrated C_3A between 300 and 400 °C corresponds to the dehydration of C_3AH_6, while the peak at about 500 °C may once again be ascribed to the disintegration of $Ca(OH)_2$. In the presence of gypsum the magnitude of the C_3AH_6-effect is drastically reduced, but instead there is a definite endopeak between 150 and 200 °C, which is associated with the dehydration of ettringite. Hydrated C_4AF behaves similarly to C_3A when subjected to heat.

The differential thermoanalysis of portland cement gives a thermogram which could be obtained by superimposing the separate effects described. From comparisons of numerous DTA curves obtained from the literature on the subject, *A. Petzold* and *I. Göhlert* [47] arrived at a DTA-curve of "highest probability" for hardened portland cement. This curve is shown in Fig. 3. It can be seen from this that the reactions associated with the giving off of water do not occur continuously, but periodically in certain temperature ranges. The effect which this may have on the structure and bonding of the cement mass is not insignificant. In point of fact microscopic observations by *K. D. Nekrasov* [6, page 62] showed that numerous cracks and fissures arise.

Little is so far known about the phase-building processes in the transient range (about 200 to 1000 °C). Experimental results are admittedly known for the separate calcium silicates from the work of *L. Heller* [48] and these have been confirmed by *O. P. Mčedlov-Petrosjan* and his collaborators [49] on the basis of thermodynamic calculations. These results show that the disintegration of hillebrandite β-C_2S results in the formation of a mixture of C_2S and SiO_2 or CS from afwillite, and also for the formation of a mixture of CS and C_2S from foshagite, β-wollastonite from xonotlite, and C_2S from okenite. It is, however, questionable whether these phases also arise in cement which has set. Investigations by *Z. Šauman* [50] and *C. D. Lawrence* [51], however, have shown that the baking of portland cement always results in the formation of wollastonite amongst other combinations. This, apart from free CaO, appears to be the only phase whose presence has been established with certainty; little is generally known about other phase formations. The occurrence of free CaO denotes a weakness in portland cement, because hydration will take place in a damp environment and as a result cracks may be formed leading to damage to the cement structure. The formation of $Ca(OH)_2$ from CaO results in a volumetric expansion of about 44%.

Strain-shrinkage phenomena are of particular importance in regard to the first heating of refractory concrete linings when they are brought into service and the subsequent first cooling down. The binding agents undergo an irreversible volumetric change as a result of the reactions described above, when they are first heated up. This change is reflected to a greater or lesser extent in the concrete itself.

Strain-shrinkage curves for pure hydrated clinker minerals have been obtained by *K. D. Nekrasov* amongst other investigators [6]. The shrinkage value corresponds basically to the loss of mass (that is, to the water given off) and lies in the range 0,5 to 0,8% up to 800 °C for all clinker minerals.

Investigations on portland cement show a similar picture; a number of different authors agree on a value of shrinkage of about 1,5% up to 900 °C. Cooling to room temperature causes a further contraction so that the total change in length after the first heating cycle has a value of about —2,4%. An example is given in Fig. 4.

When the reactions in the cement have died down after the first baking process, the cement structure exhibits a quite normal, reversible, expansion with temperature of approximately 1% at 1000 °C. The linear coefficient of thermal expansion is therefore about $100 \cdot 10^{-7}$ per °C.

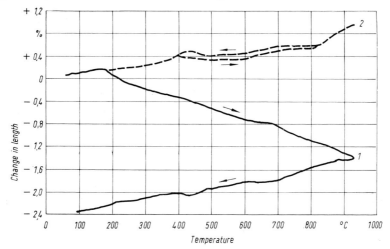

Fig. 4. Expansion/Shrinkage behaviour of portland cement
1 First temperature cycle; *2* Second temperature cycle

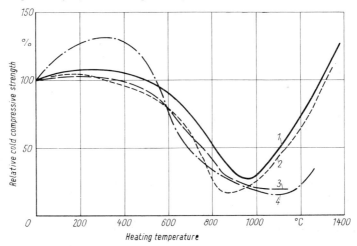

Fig. 5. The relative cold compressive strengths of portland cements after subjection to heat
1 Gurvič; *2* Braniski; *3* Sassa; *4* Ojamaa

The heating of cements to temperatures in the middle to high range results in noticeable deterioration of the strength, due to disintegrating reactions and the associated phase-building. This fact is the reason for the critical strength range of refractory concrete, which can extend over a varying region of the temperature scale, depending upon the type of cement, the additives, and so on, until the point is reached at which there is a renewed increase in strength caused by ceramic bonding.

K. D. Nekrasov and his fellow-workers are to be given credit for their fundamental investigations into the strength behaviour of different hydrated clinker minerals in relation to their previous temperature history. Numerous contributions to the subject of the temperature dependence of strength in portland cement have been made by *P. P. Budnikov* and *D. S. Il'in* [33], *I. E. Gurvič* [29], *G. M. Ruščuk* [32], *G. N. Duderov* [52], and *K. D. Nekrasov* [6]. A selection of values of strength after heating to various temperatures is shown graphically in Fig. 5. From this it can be seen that after an initial increase in strength in some cases, there is a general loss of strength from about 300 °C to as high as 1100 °C in many cases. At higher temperatures the strength again begins to increase and may sometimes reach very high values.

2.1.2. Iron Portland Cement and Blast Furnace Cement

Of the slag cements of the portland type, by which is meant those made principally or almost exclusively of blast furnace slag, the only types which need to be considered for the production of refractory concrete are iron portland cement and blast furnace cement, since the almost pure slag cements, such as supersulphated cement, are not relevant to the subject. In the German Democratic Republic the standards TGL 9272 [53] and 10 573 [45] apply to the two first-named slag cements.

According to TGL 9272 iron portland cement and blast furnace cement are defined as hydraulic binding agents. They are made from finely ground portland cement clinker, quenched blast-furnace slag and gypsum or anhydrite stone. The composition of blast-furnace slag lies within the following limits:

Blast-furnace cement, quality 225	41 to 80%
Iron portland cement, quality 225	16 to 40%
Iron portland cement, quality 350	1 to 15%

The blast furnace slags which form important constituents of the slag cements are waste products which arise during the manufacture of pig iron from the cooling of a lime-based silica smelt and which in certain conditions can be hardened by a hydraulic process. They contain essentially CaO, SiO_2, Al_2O_3 and MgO but nevertheless are poorer in lime and richer in silicic acid and alumina than portland cement clinker.

2.1.2.1. Chemical and mineral composition

The chemical composition of cement slags, expressed in terms of their most important constituents, lies approximately within the following limits:

CaO	35 to 45%	MgO	2 to 16%
SiO_2	28 to 40%	MnO	1 to 3%
Al_2O_3	8 to 18%	CaS	2 to 5%

Although MgO has a beneficial effect on the hydraulic behaviour which cannot be ignored, slags will in general be considered in the context of the three-component system $CaO-Al_2O_3-SiO_2$. This simplification is satisfactory for small MgO contents while for larger contents of MgO the four-component system $CaO-Al_2O_3-SiO_2-MgO$ must be used when considering slags.

The position of the slag area in the Rankine diagram can be seen from Fig. 6. The arrangement of the triangles of compounds shows that CS, C_2S, C_2AS (gehlenite) and possibly also C_3S_2 and CAS_2 (anorthite) would evolve as the products of crystallisation. In the presence of MgO crystallisation also results in the formation of other compounds, such as CMS (monticellite), C_2MS_2 (akermanite), C_3MS_2 (mervinite) and melilithe (the mixed crystal range gehlenite-akermanite), sometimes also to olivine (the mixed crystalline series forsterite to fayalite) pyroxene or spinell. In cement slags these phases occur in quantities so small that they may be ignored, but when heated as in refractory concrete under service conditions, they may be formed by devitrification; thus the phase equilibrium of the relevant multi-component system may be applicable to such compounds.

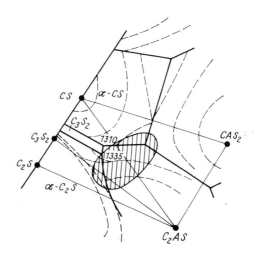

Fig. 6. The position of the slags in the $CaO—SiO_2—Al_2O_3$ system

Table 1. The chemical composition of iron portland cement (70:30) and blast-furnace cement (30:70), compared to that of portland cement clinker and blast-furnace slag

Constituents	Portland cement clinker [wt. %]	Iron portland cement [wt. %]	Blast-furnace cement [wt. %]	Blast-furnace slag [wt. %]
CaO	66	60	51	45
Si_2	22	25	30	33
Al_2O_3	6	9	12	15
MgO	2	2,5	3	3,5
SO_3	0,3	0,2	0,1	—
CaS	—	1	2	3
Alkali	1	0,5	0,2	—

The mineralogical composition of the slag cements is also a function of the ratio of clinker to slag in the same way as the chemical composition is. For a ratio of 70:30 in iron portland cement and 30:70 in blast furnace cement the analyses given in Table 1 can be obtained and compared with those of portland cement clinker and slag.

2.1.2.2. The thermal properties of hardened slag cements

The slag cements harden in a manner similar to portland cement; the gain of strength is, however, slower. An important fact which influences the behaviour of refractory concrete is that no, or very little, free $Ca(OH)_2$ occurs during

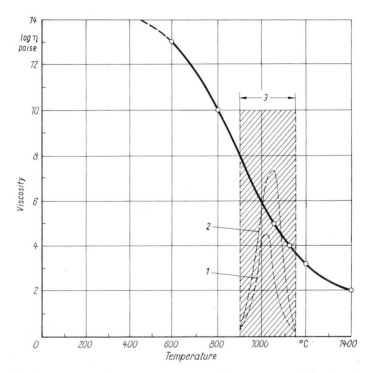

Fig. 7. Diagram showing the variation with temperature of the viscosity and crystallisation behaviour of slags (diagrammatic; after Endell and his colleagues)

1 Rate of formation of crystal nuclei; 2 Rate of crystallisation; 3 Crystallisation region

hydration. For this reason so-called ceramic stabilizers, necessary in the case of portland cement, are not necessarily required.

The simplified Rankine diagram shows that two eutectics with temperatures of 1265 and 1310 °C and a dividing point at 1335 °C lie in the portion of the diagram relating to blast furnace slag. Assuming equilibrium conditions the first appearance of melting should therefore manifest itself at these temperatures. Vitreosity of the slag and the presence of MgO pushes these stages, however, to lower temperatures. Thus, various investigations into the phase-relationships of separate part-systems of the four-component system $CaO - Al_2O_3 - MgO - SiO_2$ have shown, to quote two examples, that a eutectic occurs at $1235\,°C$ in the $CS - CAS_2 - CMS_2$ system and at $1220\,°C$ in the $S - M_2S - CAS_2$ system.

Vitreous substances have a well-known property in that, by contrast with the same substances in the crystallised phases, they have no melting point, only a region of softening. This means that when they are heated they progress steadily from the solid state through one of pasty, tough viscosity to a final liquid state of low viscosity. The commencement of softening therefore lies appreciably lower than the melting point of the crystal or crystal mixture. This is true for all glasses and also in principle for vitreous slag. Certain basic slags, however, exhibit a fairly strong tendency towards crystallisation in certain temperature ranges and this obscures the typical softening process of the glasses. The viscosity and devitrification properties of various types of slag have been studied by *K. Endell* and his fellow-workers [54] amongst others. Figure 7 serves to illustrate the character of these processes. This shows that at $1000\,°C$ relatively low viscosity is to be expected, leading to subsequent softening.

The cone fusion test point of a blast furnace cement has been determined by *H. Gibbels* [55] to have a value of approximately $1250\,°C$. The relationship between strength and temperature for slag cements is much the same as that of portland cement.

2.1.3. Pozzolanic Cement

By the term pozzolanic cement is meant mixtures of portland cement and pozzolana; these are usually silica-rich highly reactive naturally occurring or artificial substances. They have the property of reacting with the hydrate of lime split off during the setting process in portland cement and of hardening in a manner similar to hydraulic hardening. For pozzolana of this type, also sometimes known as "hydraulic aggregates", raw materials of volcanic origin, such as tuff (known in the ground-up state as trass) and santorin are used, as also materials of marine origin, such as kieselguhr, diatomite, moler, and *Tripoli powder*, and sometimes also artificial products, for instance silica dust, flue-dust and fly-ash.

In many countries pozzolanic cements are used on a large scale and in some cases even standardised. Their technical and thermal properties are similar to those of blast furnace cements.

2.1.4. Aluminous Cement

Three types of aluminous cement may be distinguished: normal industrial aluminous cement, aluminous cement of high alumina content, and barium aluminous cement. The first-named is by far the most common. It is often

3 Petzold/Röhrs

also described as "smelted aluminous cement"[1][*]. This normal industrial aluminous cement will be dealt with first.

Aluminous cement is described in TGL 9738 [58] as an hydraulic binding agent, consisting of compounds of calcium oxide and aluminium oxide together with small quantities of silicon dioxide, iron oxide, titanium oxide, and magnesium oxide. It is obtained by the smelting of limestone or quicklime together with bauxite or other materials having a high alumina content, in an electric furnace, followed finally by grinding of the cooled clinker without any further additives.

2.1.4.1. Chemical and mineralogical composition

The chemical composition of the well-known industrial or commercial aluminous cements is found to lie within the following limits:

CaO	32 to 44%	FeO, Fe_2O_3	1 to 18%[2]
Al_2O_3	35 to 50%	MgO	0.2 to 1%
SiO_2	4 to 14%	TiO_2	1 to 2%

The variations in proportions are therefore much greater than in portland cement. The following rule of thumb is useful for the main constituents of "smelted aluminous cement":

CaO	approx. 40%	SiO_2	approx. 10%
Al_2O_3	approx. 40%	FeO, Fe_2O_3	approx. 10%

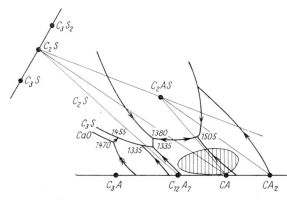

Fig. 8. The position of aluminous cement in the $CaO-SiO_2-Al_2O_3$ system

Aluminous cement normally signifies a pure calcium aluminate cement. It is described, with reference to its phase relationships, however, in terms of the $CaO-Al_2O_3-SiO_2$ system, on account of its consistently high SiO_2-content. The place occupied by aluminous cement in this system is shown in Figure 8.

---

[1]) According to American investigations, aluminous cement produced by an electrical smelting process markes for better refractory concrete than aluminous cement produced by other methods [56]. This view is not shared, however, by Hungarian experts [57].

[*]) *Translator's note:* "smelted aluminous cement" is an approximate translation of the German term "Tonerdeschmelzzement".

[2]) According to refractory experts, the Fe_2O_3-content permitted by the TGL standards to be 10%, should be reduced to not more than 4%, if the "smelted aluminous cement" is to be used for refractory concrete.

From the position of the "smelted aluminous cement" region and the relevant triangles of compounds in this diagram, the following combinations can be obtained for the separate phases:

a) $CA - C_{12}A_7 - C_2S$ ("First quality smelted aluminous cement"),
b) $CA - CA_2 - C_2AS$ ("Second quality smelted aluminous cement"),
c) $CA - C_2S - C_2AS$ (substandard, inferior).

It is recognised that commercial aluminous cement is grouped around the monocalcium aluminate CA, whose principal component distinguishes it. In addition to the compounds described above, the following can also be found in measurable quantities: C_2F, C_4AF, MA, C_6A_4MS, $C_6A_4F\ddot{\,}S$, $CaO \cdot TiO_2$ and Fe_3O_4.

2.1.4.2. Hardening of aluminous cement

The characteristic hardening process and the phase-building, which are determining factors in the behaviour of a cement when heated, were a major reason why for a long time aluminous cement was used almost exclusively in refractory concretes.

In contrast to the C_3A contained in portland cement, the aluminates of low lime content, particularly CA and CA_2, exhibit very valuable characteristics when hardening. The most evident is the high initial strength which distinguishes "smelted aluminous cement". Hydration takes place, not by the separation of lime, but by the formation of $Al(OH)_3$ and lime-rich hydroaluminates. The separated $Al(OH)_3$ is amorphous in the first place, then adheres to the other particles and subsequently changes into crystalline hydrargillite. The other clinker minerals, such as C_2S and C_4AF, hydrate in the well-known manner. Hardening takes place accompanied by the evolution of a large amount of heat, due to the high heat of hydration of the aluminate and the concentration of the released heat as the result of rapid hydration. Hardening is considerably influenced by temperature; low temperatures are advantageous, while above 25 °C the strength deteriorates appreciably.

2.1.4.3. Thermal properties of the hardened cement structure

If aluminous cement consisted of pure calcium aluminates, then it would be expected from the diagram for the $CaO{-}Al_2O_3$ system given in Fig. 9, that on the lime-rich side a minimum softening temperature of 1400 °C would obtain, and on the aluminate-rich side a minimum of 1590 °C (eutectics $C_{12}A_7 - CA$ and $CA - CA_2$ respectively). In the ternary system aluminous cement of the first quality has a eutectic which forms at 1335 °C and aluminous cement of the second quality a eutectic forming at 1505 °C. Admixtures of iron oxide (bivalent and tervalent) and magnesia cause these temperatures to drop still further. Effective reduction during smelting can so reduce the iron content that the fire-resistance of the cement is improved by at least 30 °C.

The considerable variations found in the chemical composition of aluminous cements usually result in varying values of the cone fusion test temperature. Thus the cone fusion test point for normal aluminous cements with a low iron content varies from 1320 to 1430 °C, [11], [12], [17], [13], [59], while that for iron-rich cements lies between 1230 and 1270 °C [12], [13], [59].

Investigations by Soviet scientists into the compressive resistance at high temperature of hydrated calcium aluminate have shown that softening commences at 1250 °C in $C_{12}A_7$ and a 4% deformation occurs at 1350 °C. For CA and CA_2 the temperatures were 1360 and 1450 °C respectively.

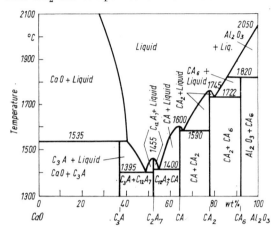

Fig. 9. The $CaO-Al_2O_3$ system

2.1.4.4. Alterations with temperature

Aluminous cement, like all other set cements, loses its adsorbed and chemically combined water upon being subjected to heat. A number of thermoanalytical studies on hydrated aluminates of low lime-content have shown in essence only dehydrating effects, which extend their influence to about 300 °C and cause a partial and gradual loss of water (Fig. 10).

Similar thermograms can be obtained for hydrated aluminous cement. The differential thermoanalysis curve derived for smelted aluminous cement by Zschornewitz is also given in Fig. 10; it agrees in its essentials with the curves published by other authors. The complex endothermic effect between room temperature and about 300 °C denotes the loss of adsorbed water and the dehydration of calcium aluminate hydrate and aluminium hydroxide or aluminium oxide hydrate. The dehydration reactions of hydroaluminate should be completed by 400 to 500 °C at the highest. Since it is evident that the rejection of water takes place gradually, the term "zeolitically combined water" is sometimes used in this connection (*Budnikov, Nekrasov*). Effects which appear later can presumably be ascribed to other hydrate phases.

With regard to phase-building under the effect of temperature it can be deduced from studies by *H. Mitusch* [12], that γ-alumina can occur temporarily above 500 °C, and CA_2 above about 1000 °C. *S. J. Schneider* [59] has shown that in cements which have a low iron content, at a temperature of about 900 °C, CA, C_2AS, $CaTiO_3$, $C_{12}A_7$ and Fe_3O_4 are produced; these phases, with the exception of $C_{12}A_7$, are retained even at 1400 °C. In iron-rich cements, at 900 °C, CA, $C_{12}A_7$, C_2AS, C_2F and $CaTiO_3$ are formed; in this case also $C_{12}A_7$ disappears finally at 1300 °C.

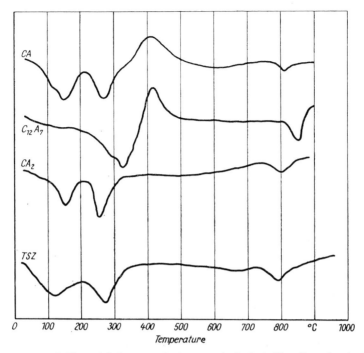

Fig. 10. Differential thermoanalysis curves for hydrated low-lime aluminates and for aluminous cement (TSZ = Aluminous cement)

Fig. 11. Expansion/Shrink-age behaviour of aluminous cement having high and low contents of iron

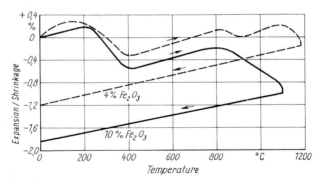

Irreversible shrinkage takes place in parallel with the reactions of dehydration. The character of the shrinkage is very similar for the separate aluminates and for their hydrated phases. Typical strain-shrinkage curves for aluminous cements of high and low iron contents are given in Fig. 11 [12], [59], [60]. The occurrence should be noted of a maximum point at 800 °C (and again in one of the cases at about 1100 °C) which arises from the intermediate expansion of the completely dehydrated cement structure, before a second stage of shrink-age, which is evidently associated with sintering, takes place. The contraction of the high-iron cement up to 1000 °C has a value of 0,6 to 1,0%; the test specimen contracts about a further 1% on cooling so that there is a total shrinkage after one heating cycle of about 2%. The behaviour of the low-iron aluminous cement is very surprising. The maximum total shrinkage in this case is only about 1,2%, which must be considered as a great advantage for use in refractory concrete; the small shrinkage which would take place in concrete could well be ignored.

Subsequent reheating of aluminous cement causes the normal reversible thermal expansion. The average value of the coefficient of expansion up to 900 °C is 60 to 80 · 10^{-7} per °C. Above 1000 °C the stage of irreversible shrinkage as a result of sintering has been reached.

The only relevant information about the effect of temperature upon strength for pure hydrated calcium aluminates comes from the work of *K. D. Nekrasov* [6, page 150]. The loss of strength in actual aluminous cements extends over a wide range of temperature, usually between 100 and 1000 °C, but the minimum value of strength is usually found at about 900 °C; this critical temperature can be either higher or lower, however, depending upon the nature of the cement; the amount of the loss of strength may also vary considerably, from 40 to 80%.

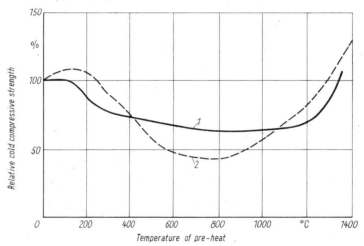

Fig. 12. Relative value of cold compressive strength of aluminous cements after subjection to heat
(*1* Nekrasov, *2* Braniski)

Typical strength curves for pure aluminous cement are given in Fig. 12 [6], [13]. An interesting point — previously overlooked — is that occasionally a minimum strength is reached at temperatures as low as 50 to 100 °C; this phenomenon can be attributed to the partial dehydration of the CAH_{10} and the formation of C_3AH_6 and $Al(OH)_3$. The immediate subsequent increase in strength is presumably a result of the increase in the formation of the latter phase and of accelerated crystallation [59]. The modulus of elasticity also behaves in a similar manner.

2.1.5. High Alumina Cement

While the content of Al_2O_3 in normal industrial "smelted aluminous cements" does not exceed 50% and these cements usually fall therefore in the triangle of compounds of the first quality cement with a eutectic at 1335 °C, special aluminous cements with a high Al_2O_3 content are also produced. These have much better fire-resistant characteristics and are therefore used for special refractory concretes.

2.1.5.1. Chemical and mineralogical composition

The alumina content of these cements must be 64 or 65% or greater, depending upon the composition of the low-lime aluminates, if CA or CA_2 are the principal constituents. A selection of analyses of special cements of this type is given in Table 2.

Table 2. The chemical composition of high-alumina cements
(from various sources of information)

Constituents	France SECAR [12] [wt. %]	Germany Rolands- hütte Super [12] [wt. %]	CA-cement [13] [wt. %]	CA_2-cement [13] [wt. %]	USA [59] [wt. %]	France SECAR 250 [63] [wt. %]
Al_2O_3	64,2	68,6	62,8	72,7	81,3	70 ···72
CaO	34,1	24,9	35,6	26,5	17,4	26 ···29
SiO_2	0,4	2,9	0,5	0,3	—	0,5··· 1
Fe_2O_3	0,3	0,4	0,4	0,2	0,3	0,5··· 1
MgO	0,6	0,5	0,6	0,3	0,6	—

Because of their purity these cements may be considered in terms of the two-component system, $CaO-Al_2O_3$. As may be seen from Fig. 9, the low--alumina combinations lie near the determining point of CA (theoretically 33,5% of CaO and 64,5% of Al_2O_3) and, in accordance with the laws of heterogeneous equilibrium, consist predominantly of this mineral. The high-alumina cements lie between the determining points of CA and CA_2 (the latter being theoretically 21,6% CaO and 78,4% Al_2O_3) and sometimes also between CA_2 and CA_6 and therefore consist of a combination of these phases.

2.1.5.2. Thermal properties and behaviour when heated

The temperatures at which the first smelting occurs can be seen from the arrangement of the eutectics in the $CaO-Al_2O_3$ system. For cements with a content of Al_2O_3 of less than 64,5%, softening in the eutectic $C_{12}A_7-CA$ begins at about 1400 °C; however, since this eutectic is very close to the determining point of $C_{12}A_7$ and also the alumina contents of the cements are not very different from the composition of CA, the small quantities which commence to smelt do not significantly influence the fire-resistance. The actual values given for the cone fusion test point of "white" cements of this type by *J. Arnould* [63],

A. *Braniski* [13], and *H. Mitusch* [12], are from 1520 or 1530 to 1580 °C. The presence of an appreciable quantity of SiO_2 causes the composition to move into the conjugation triangle of second-quality smelted aluminous cement in the Rankine diagram, with a eutectic temperature of 1505 °C, which compares well with the cone fusion test values mentioned. Good fire-resistance properties can therefore be attained in the ternary system as well so long as the composition lies in the correct triangle.

In cements with an Al_2O_3 content greater than 64,5% softening is determined by the eutectic CA -- CA_2 at about 1590 °C. High quality cements with an Al_2O_3 content of about 73% therefore do not begin to soften until an advanced stage. *J. Arnould* [63] has given a cone fusion test point of 1580 °C for a cement of this type and *A. Braniski* [13] a figure of 1610 °C. *S. J. Schneider* [59] has stated a cone fusion test temperature of 1750 °C for a high-alumina cement having an Al_2O_3 content of 81%, a figure which agrees with the phase diagram.

Resistance to compressive stress at high temperature is greatest in cements which have an Al_2O_3 content greater than 65%. Values of t_a (4% deformation) have been found to lie in the range 1450 to 1520 °C [60].

The hardening process in aluminous cements with a high-alumina content is similar in principle to that for normal commercial smelted aluminous cement. It is therefore to be expected that the cement structure will show similar chemical effects and processes when subjected to heat. *S. J. Schneider* [59] has shown that the DTA-curve for a high-alumina cement with a content of 81% Al_2O_3 is always very similar to that for a normal aluminous cement. Radiographic examination has shown that the hydrated phases in a set cement transform at 900 °C into Al_2O_3, CA, CA_2 and $C_{12}A_7$ and that at 1400 °C only the presence of CA_2 can be demonstrated.

G. V. Kukolev and *A. I. Rojzen* [60] have shown that the shrinkage behaviour when the cement is subjected to heat is dependent upon the chemical composition. The smallest shrinkage values (0,3 to 0,5% between 100 and 800 °C) are found in cements with an Al_2O_3 content of 70%. High-alumina cement with an Al_2O_3 content of 81% was found by *S. J. Schneider* to exhibit a noticeable irreversible expansion of about 2%, from about 1000 °C, which he attributes to the formation of CA_2. In this case, of course, the total resultant effect is a nett expansion, not shrinkage, of about 0,5% in the test specimen.

The coefficients of expansion of CA_2-rich cements can be stated to be from 34 to 67 $\times 10^{-7}$ per °C and those of CA cements from 83 to 100 $\times 10^{-7}$ per °C.

The strengths of high-alumina cements do not decrease appreciably upon heating, according to *Kukolev* and *Rojzen*, a feature which is well known in normal smelted aluminous cement. Investigations by *Schneider*, however, have shown that high-alumina cement loses strength as much as all other cements and in other respects behaves like normal aluminous cements, with regard to the curves of strength and modulus of elasticity.

2.1.6.   Barium Aluminous Cement

The development of barium and strontium cements is mainly due to *A. Braniski*, who has researched into and described the properties of these siliceous and aluminous binding agents; *P. P. Budnikov* and his colleagues stand out among later workers in the field of barium aluminous cements. Barium aluminous cement based upon monobarium aluminate (BA)[1] is particularly to be recommended for fire-resistant concretes, on account of its especially good fire-resistant properties [13], [64], [65], [66].

2.1.6.1. Chemical and mineralogical composition

The investigations of Soviet experts have been particularly directed towards pure monobarium aluminate with a theoretical composition of 62% BaO and 38% Al_2O_3 [65], [67], [68], [69]. *A. Braniski*, however, has worked on technical (commercial) barium aluminous cements, having compositions in the following range:

BaO	58	to 63%	Fe_2O_3	0,3%
Al_2O_3	39	to 29%	MgO	0,1%
SiO_2	0,7	to 6%	Alkali	0,2%
CaO	0,3%			

The smelting and phase equilibrium characteristics can be seen from the $BaO-Al_2O_3$ system in Fig. 13. This shows that BA predominates in these cements with small quantities of B_3A or BA_6. In the presence of SiO_2, BAS_2 (barium felspar, celsian) presumably occurs as a ternary phase in the $BaO-Al_2O_3-SiO_2$ system. This system will not be illustrated here.

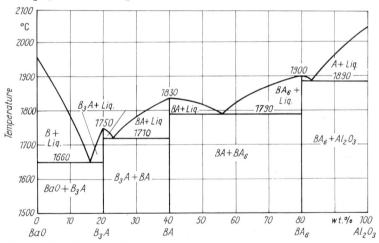

Fig. 13. The $BaO-Al_2O_3$ system

2.1.6.2. Thermal properties

The smelting behaviour of barium aluminate can be seen from the phase-relationship diagram. The very high melting-points of the pure compounds should particularly be noted (B_3A 1750 °C, BA 1830 °C, BA_6 1900 °C) as also

---

[1] According to the work of *A. Braniski* [64] barium aluminous cement is relatively water-soluble by comparison with other cements, and in the strictest sense is therefore an air-binding agent rather than an hydraulic binding agent. It is considered in the cement group or family because of the close analogy between it and the lime-aluminous cements.

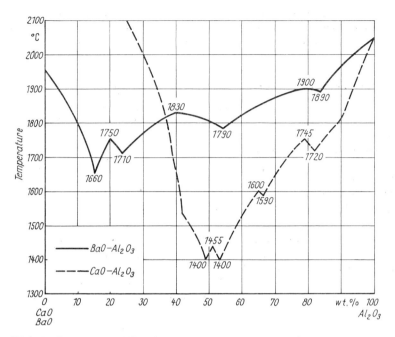

Fig. 14. Comparison of liquid temperatures in the $CaO-Al_2O_3$
and $BaO-Al_2O_3$ systems

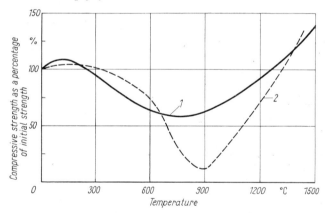

Fig. 15. Relative values of cold compressive strength of barium
aluminous cement after subjection to heat, compared with those
of portland cement (Braniski)
1 Barium aluminous cement; 2 Portland cement

the relatively flat eutectics which lie near to these temperatures. The particular
advantage over the lime-aluminates can be seen from the comparison of the
liquid temperatures of the two systems given in Fig. 14. The fire-resistance of

the material is very much influenced by silica, iron oxide and alkali. *Braniski* gives a cone fusion test point temperature of 1730 °C for a cement containing silica, but 1810 °C for a high quality cement.

Upon heating of the hardened cement the hydrates decompose, so that water must be given off continuously and in quantity (69). The strength decreases with increase in the temperature of preheat until at 600 to 700 °C it has fallen to about 60% of the initial strength. A curve indicating this is shown in comparison with the corresponding curve for portland cement in Fig. 15 (after *A. Braniski* [64]). The relatively high compressive strength retained by barium aluminous cement in the critical region, when compared with other cements, is particularly to be noted.

2.2. Anhydrous, Inorganic, Cold-hardening, Binding Agents

As already mentioned in the introduction to this chapter, the development of high-temperature concretes has extended beyond those which have a cement basis to the manufacture, testing and use on a commercial scale of concrete-type materials using other binding agents. These agents are inorganic and anhydrous and set without the application of heat. In the Soviet Union, for example, materials using waterglass and magnesia binding agents have been developed to a considerable extent, and in the United States phosphate-bound substances. A brief discussion of binding agents of these types now follows.

2.2.1. Waterglass

The term waterglass denotes silicates of sodium or potassium of various compositions, which have been dissolved or colloidally mixed in water and are presumed to be almost completely hydrolised. The most commonly used waterglass is a silica-rich sodium silicate, containing from 2 to 4 gramme molecules of SiO_2 to 1 gramme molecule of sodium silicate. Hardening takes place in air, usually with the use of hardening accelerators as additives.

Detailed information about the chemical problems associated with waterglass can be found in the work of *B. Butterling* [70] amongst others.

2.2.1.1. Composition and properties of a waterglass solution

The ratio $SiO_2:Na_2O$ is a deciding factor in the usefulness of waterglass as a binding agent for waterglass refractory concrete.[1] Experience has shown that this ratio (known as the Waterglass Ratio) should have a value of between 2,4 and 3,0 [42]. Industrial waterglass is a colloidal solution of highly hydrated silica, which is dispersed by the presence of hydrolytic sodium hydroxide. There also exists in the solution the compound $Na_2Si_2O_5 \cdot 9H_2O$, as *N. S. Dombrovskaya* and *M. R. Mitel'man* have shown [71].

Waterglass solutions are made having specific gravities ranging from 1,32 to 1,40. Soviet sources state that the density of the solution for waterglass

---

[1]) In the text which follows, all concrete-like materials, even though they are not cement-bonded, will be referred to as high-temperature concretes, refractory concretes, or fire-resistant concretes, for the sake of simplicity.

concrete should be from 1,38 to 1,40. In this form they contain approximately 27% SiO_2 and 8% NaOH [72].

The chemistry of waterglass hardening results from the complete hydrolysis of the dissolved sodium silicate under the action of the carbon dioxide in the atmosphere, resulting in the separation of silica in the form of a gel, which then sets by condensation into a polysilica structure. A number of different substances may be used as accelerators (for example, silicofluoride, sulphide, organic substances) but of these only sodium silicofluoride Na_2SiF_6 has proved to be of practical use on a large scale. In the Soviet Union all waterglass refractory concretes are therefore made using Na_2SiF_6 exclusively as the additive. The most suitable percentage has been found to lie in the range 10 to 15% of the waterglass quantity.[1] Reference should be made to the specialist literature for the use of these additives. In the presence of aggregates reactions may take place which, even at room temperature, complicate the hardening process.

2.2.1.2. Properties of hardened waterglass

In comparison with the true cements, waterglass has no advantages in itself, nor in the pure, hardened structure made from it, other than in respect of modulus of elasticity and density. The only points of interest, therefore, lie in the thermal characteristics which may be obtained by the use of waterglass concrete-type materials.

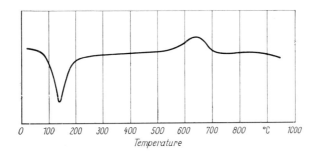

Fig. 16. DTA curve for hardened waterglass

The differential thermoanalysis curve in Fig. 16 indicates that extensive dehydration of the silica gel occurs between the temperatures of 100 and 150 °C, resulting in a water-free gel mass of vitreous appearance. The slight exopeak in the 600 to 650 °C temperature range can be attributed to crystallisation phenomena of the $Na_2O \cdot 2SiO_2$. No other phase-building has been observed.

The low smelting and eutectic temperatures of the sodium silicates may seem surprising for materials to be used in the manufacture of temperature-resistant materials. The phase-diagram for $Na_2O - SiO_2$ given in Fig. 17 shows that $Na_2O \cdot 2SiO_2$ melts at 874 °C, and that combinations of this silicate with SiO_2, which lie in the waterglass range, appear as a eutectic at 790 °C. Their use-

---

[1] Following the proof by *Ja. V. Ključarov* and *N. V. Mesalkina*[73] of the action of calcium oxide as a stimulating agent in magnesia waterglass concretes, *N. V. Jl'ina* and *L. I. Skoblo* [74] have recently been able to demonstrate that the hardening of waterglass is accelerated by even a small percentage of portland cement.

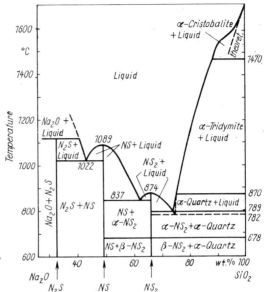

Fig. 17. The Na_2O-SiO_2 system

fulness as binding agents in waterglass refractory concrete lies in their high reactivity, which enables reactions to take place with a great variety of aggregates, resulting in compounds which are thermally stable and have good fire-resistance.

2.2.2. Magnesia Binding Agents

Magnesia binders are produced from caustic magnesia (lightly burnt, unsintered MgO) and magnesium chloride or magnesium sulphate solutions. The caustic magnesia can be obtained either by the burning of magnesite or as a by-product of the manufacture of magnesium sulphate. The technical requirements for magnesia to be supplied for civil engineering purposes are given in TGL 12375 [75]. They are, however, not directly applicable to its use in concrete-type materials for high-temperature purposes, though some of the requirements and properties are the same for both cases.

The caustic magnesia should be present in the form of a finely-ground dust, so that a satisfactory reaction with the chloride or sulphate solutions may be ensured (retention on a 0,09 mm sieve should be less than 25%). If the fire-resistance is to be assured, the content of impurities must be limited, but exact figures are not known.

There are also no exact requirements expressed for the $MgCl_2$ or $MgSO_4$ solutions which are used as reactants. *A. B. Searle* [76] gives a density of 1,12 g/ml for the $MgCl_2$ solution for a magnesia cement. A number of Soviet papers mention densities of 1,20 to 1,22 g/ml for $MgSO_4$ solutions.

The information about mix proportions in magnesia binding agents of this type is equally scarce. In agents of the Sorel's cement type for general civil engineering purposes the ratio normally used for $MgO:MgCl_2$ is 2,5:1. In the

Soviet Union concrete-type linings for foundry ladles have been constructed using an addition of 4 to 5% of magnesium sulphate solution to the total mix (MgO, chrome ore and broken chrome ore-magnesia) [77].

There are still many gaps in the scientific knowledge about the processes which take place during the hardening of magnesia binding agents. It is assumed that hardening is caused by a flocculation of colloidal $Mg(OH)_2$, which thus bonds with the aggregate. In addition, magnesium oxychloride occurs, leading to the formation of a secondary crystal structure and so to hardening.

The thermal behaviour of magnesia binding agents, in so far as it is capable of being observed, has not yet been the subject of separate study. Only isolated observations have been made of it in association with aggregates. Fundamental alterations during heating are not to be expected. Internal reactions within the binder will not occur other than dehydration and the expulsion of chlorine or SO_2 at very high temperatures. With the crystallisation of periclase (native magnesia) reactions with certain aggregate materials are the only processes that appear possible.

It should be mentioned here that in practice (notably in the Soviet Union) it is not general to use specially prepared caustic magnesia. There is ample evidence to show that finely ground sintered magnesia can be used in its place with success (particle size less than 0,09 mm); in general the finest fraction of the magnesia aggregate is simply allowed to act as the binding agent. These binding agents are most commonly described as "Periclase cement".

2.2.3. Dolomite

It is well known that dolomite, if burnt caustically under certain conditions, can possess hydraulic binding properties. This has also been shown by *P. P. Budnikov* to be true for a well burnt and subsequently finely ground mixture of dolomite, chromite and quartzite [10]; by this means he has produced hydraulically-bound materials having a 28-day compressive strength of 270 kp/cm^2 (3830 psi), a t_a-value (defined in Section 2.1.5.2) of more than 1500 °C, and possessing good stability under abrupt changes of temperature.

This type of binding agent does not, however, appear to be of great commercial importance.

2.2.4. Phosphoric Acid and Phosphates

The possibility of making tamped mortars and refractory bricks by the use of phosphoric acid or phosphate has been known for about 40 years, but has only come to be used on a commercial scale in the last 15 years [78], [79], [80], [81], [82], [83].

In addition to orthophosphoric acid, H_3PO_4 acidic aluminium phosphate, for which the formula is $Al(H_2PO_4)_3$, also merits a few observations[1] (mono-aluminiumdihydrogenphosphate). It is the basis of a number of commercial

---

[1] The suitability of phosphoric acid or phosphate depends upon a number of factors, among them the type of aggregates to be used. For example, H_3PO_4 is preferable in conjunction with alumina or corundum, while phosphate is more suitable in association with materials containing zircon [86].

binding agents, e.g. the material described by *H. Bechtel* and *G. Plosa* [84], [85] as "Feuerfestbinder 32" for which the following details have been supplied:

Density	1,475 g/ml	P_2O_5	33,2%
$Al(H_2PO_4)_3$	46,5%	$Al_2O_3 : P_2O_5$	1:3,2
Al_2O_3	7,5%		

Phosphates or phosphoric acid are used in association with the normal fire-resistant aggregates. The mechanism of hardening is based upon a chemical exchange, which has as its base a type of neutralisation process and the formation of tertiary phosphates.

It is assumed that the process of bonding in the aluminous concretes (90% Al_2O_3, 10% H_3PO_4) described by *W. H. Gitzen, L. D. Hart* and *G. MacZura* [82], which does not occur before a temperature of 300 or 400 °C is reached, takes place as a result of the dehydration of orthophosphoric acid to pyrophosphoric acid, $H_4P_2O_7$, and its reaction with Al_2O_3 as $AlPO_4$ is produced. With other aggregates, such as MgO, the corresponding phosphates are produced, such as $Mg_2P_2O_7$ ($= 2MgO \cdot P_2O_5$) or $Mg_3P_2O_8$ ($= 3MgO \cdot P_2O_5$) [88]. Similar reactions are to be expected in connection with monoaluminiumhydrogenphosphate and here the occurrence of condensed phosphates in addition has been described [84]. The binding reactions can be assisted by the addition of clay to the mixes; clay also acts as a plasticising agent[1]).

Thermal changes to the orthophosphate are not to be expected during service unless the temperatures are very high. The melting point of $AlPO_4$ is about 2000 °C; from 1600 °C, P_2O_5 is formed and then vaporised; the remaining alumina is present in a modification similar to γ-Al_2O_3 or as α-Al_2O_3 [89], having, in this form also, the properties of a binding agent [82].

It should also be mentioned that concrete-type materials and mortars can also be made from monomagnesiumdihydrogenphosphate, $Mg(H_2PO_4)_2$ [81].

2.3. Aggregates

All naturally occurring sands and gravels may be used in the manufacture of normal structural concrete, as well as suitable prepared volcanic rock, limestone or sandstone (in the case of dense concretes) and foamed slag, slate, vermiculite, pearlite and clays (such as ceramsite) or sintered ash for lightweight concretes. For concretes intended for use at high temperatures, on the other hand, the choice of aggregates must be limited to those that are fire-resistant; fire-resistant materials can be dispensed with only in structures which have a very light thermal duty.

A general summary can and should deal only with the most important of the aggregates used in the making of refractory concrete, highlighting the principal points to which attention should be paid in this context. The thermal properties of high-temperature concretes are very largely determined by the

---

[1]) *M. A. Matveev* and *A. I. Rabuchin* [87] have described systematic research into phosphoric acid and aluminium phosphate binding agents under the most varied conditions and have particularly discussed the suitability of these mixes as high-temperature binders for ceramics and metals.

choice of aggregates; it is thus possible by the choice of suitable fire-resistant materials to produce concretes of high thermal resistance. The problems of size and grading are discussed in Chapter 3.

In the discussion which follows on the subject of aggregates in the context of the technology of refractory concrete, only the coarser materials having a particle-size greater than 0,2 mm (0,008 in) will be considered. The fine aggregates, comprising for the most part particles smaller than 0,1 mm (0,004 in), are used as stabilising agents with special chemical functions, especially in refractory concretes made with portland cement, while in waterglass concretes they act as micro-filling material. They are therefore not aggregates in the true sense, but rather additives. They will be defined and discussed as additives.

Aggregates include naturally occurring materials, although only a limited number of these are suitable. They also comprise thermally pre-treated aggregates, i.e. materials which have been previously burnt at high temperatures, although not in their final size, which may then be classified as suitable for prepared (crushed) aggregates. In this work fire-resistant materials will be classified and described generally in accordance with the rules of the refractory industries, as laid down in TGL 9397 and 99-13 [90].

2.3.1. Non-fire-Resistant Mineral Aggregates

Non-fire-resistant mineral aggregates are suitable for use only in heat-resistant concretes; they are not suitable for fire-resistant concretes. Their resistance to heat is adequate only for those concretes which are specifically intended for use at temperatures lower than 1000 or 1100 °C. In certain circumstances, however, additives which are not fire-resistant, such as slags, may also be used for high-temperature concretes, if they produce a particular desired chemical effect.

2.3.1.1. Naturally occurring materials

The naturally occurring stones which are preferred for aggregates are those with a coefficient of expansion which is not excessive and which do not undergo a volumetric increase under refractory conditions caused by irreversible modifications, such as may be expected in the case of unburnt quartzsand, gravel or sandstone. Care should be taken in the use of limestone, since variations in properties and deterioration may take place.

According to information supplied by *G. D. Salmanov* [91], diabases and basalts exhibit relatively small thermal expansion. Since they do not undergo alteration, therefore, they are equally useful, as aggregates, as pumice and andesite. Certain tuffs and porphyries are also suitable [92], [93], [17].

Moler is sometimes used in the making of lightweight refractory concrete. This is a type of kieselguhr, containing fairly large quantities of clay or volcanic ash as impurities. Its composition when fused together, comprises approximately 78% SiO_2, 11% Al_2O_3, 6% Fe_2O_3, and 4% alkaline earth. Its cone fusion test point is only 1410 °C. Calcined diatomaceous earth may also be used as a lightweight aggregate [94].

2.3.1.2.   Industrial wastes and by-products

Slags obtained from various sources are the first by-products to be considered. A number of Soviet research workers (see, for instance ref. [95]) have tested a large range of slags to ascertain their effectiveness in heat-resistant concretes up to operating temperatures of 800 °C, with the result that they are now approved constituents of heat-resistant concretes [8]. It is a requirement that the CaO-content should be less than 45% and the basicity level, as expressed by the ratio

$$\frac{CaO + MgO}{SiO_2 + Al_2O_3}$$

should be less than unity. The coefficient of thermal expansion in the range 20 to 800 °C is stated to be approximately 60×10^{-7} per °C but will vary somewhat according to the chemical composition.

Another type of waste product is broken brick. In the U.S.S.R. it is used principally in the finely ground state as a ceramic stabiliser or additive. In the form of a coarse aggregate broken brick has been successfully used for making heat-resistant concrete for tunnel furnace wagons; see the work of *M. Röhrs* [96], [97]. Detailed technical information for this material is not available.

2.3.1.3.   Artificial aggregates

The most common artificial materials which are made for use as concrete aggregates are certain lightweight aggregates intended for the manufacture of lightweight concrete. Examples are foamed clay and foamed slate, known by names such as "Ceramsite" and "Globulite". In the Soviet Union these aggregates also have been tested for use in lightweight refractory concrete. Ceramsite, with a density between 500 and 600 kg/m³ (31 to 37,5 lb/ft³) should be used to make a concrete having a density of 1400 kg/m³ (81 lb/ft³), and Ceramsite with a density of 350 to 450 kg/m³ (22 and 28 lb/ft³) for a concrete of 900 kg/m³ (56 lb/ft³) according to the recommendations of the Soviet Standards [8].

Other types of lightweight aggregates include expanded vermiculite and pearlite. According to the Soviet Standard [8], vermiculite can be used in the manufacture of waterglass-based lightweight refractory concrete.

Information about the use of foamed pearlite can be obtained from the work of *H. Mitusch* [98]. The use of foamed slate is also possible [99]. Nothing is known of the usefulness of ash sinter, such as Aggloporite, in refractory concretes, although these materials should in principle be suitable for the purpose.

The use of slag (mineral) wool has recently been recommended for heat-resistant lightweight concretes [100], in combination with powdered chamotte (sintered fireclay) and portland cement or waterglass.

2.3.2.   Chamotte

Fire-resistant chamottes[1] of all types are the most commonly used aggregates for heat-resistant and fire-resistant concretes. They are used to a considerable

---

[1] Only fire-resistant chamottes will be considered here. Non-fire-resistant materials, such as potters' clay and roken brick, will not be discussed.

extent also as additives for portland cement and waterglass cement concretes. In most cases they are utilised in a manner similar to ordinary aggregates with a suitable grading. The use of lightweight chamottes for lightweight refractory concretes has also been recorded [98].

Chamotte is made by burning, at high temperatures, naturally occurring fire-resistant kaolin rock or argillaceous rock; the principal components of the resultant material are SiO_2 and Al_2O_3, small quantities of Fe_2O_3 and alkali usually being present as well. The proportions of the constituents may vary considerably depending upon the composition of the raw aggregates.

A distinction is drawn between the quartz chamottes, or acid chamottes, in which the Al_2O_3-content is less than 30%, and the SiO_2-content lies in the range 65 to 85%, and the true chamottes having an average alumina content of 30 to 45%. The following sub-divisions are specified for commercial purposes in TGL 4323 [101]: Class I (42···45% Al_2O_3), Class II (37···42% Al_2O_3), Class III (30···37% Al_2O_3) and Class IV (68···85% SiO_2).

2.3.2.1. Chemical and mineralogical composition

In chamottes having average aluminium oxide contents, the principal constituents lie within the following limits:

$Al_2O_3 + TiO_2$	15 to 45%	Alkaline earth	1 to 2%
SiO_2	50 to 85%	Alkali	1 to 3%
Fe_2O_3	1 to 4%		

The mineralogical composition of the chamottes can be obtained only approximately from the equilibrium diagram in Fig. 18, since vitreous phases close

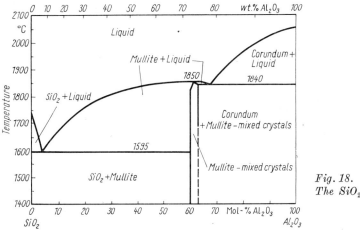

Fig. 18.
The $SiO_2 - Al_2O_3$ system

to mullite and SiO_2 (cristobalite) continually occur, due to the presence of fluxes such as Fe_2O_3, alkali and alkaline earth. In acidic chamottes a proportion of the free SiO_2 is also present in the form of unaltered quartz.

2.3.2.2. Thermal properties

The temperature of first melting in the $SiO_2 - Al_2O_3$ system is that of the binary eutectic, i.e. 1595 °C; in reality softening occurs appreciably earlier, due to the content of vitreous material, but is not noticeable until there is a continuous application of load. In general the service temperatures of the acidic chamottes are definitely below those of the others. The former are therefore not often considered as aggregates for fire-resistant concretes.

The cone fusion test point varies, depending upon the composition. Temperatures of 1630 to 1670 °C can be expected for quartz chamottes, and 1710 to 1770 °C for normal chamottes.

Softening under pressure begins relatively early in the chamottes, and a fairly wide softening range is one of their common characteristics. The t_a-values lie between 1250 and 1300 °C for the quartz chamottes, and those for normal chamottes between 1400 and 1450 °C. The corresponding t_e-values are approximately 1450 °C and from 1500 to 1600 °C respectively. The service temperatures rise similarly from about 1200 to 1400 °C with increasing alumina content.

Reversible thermal expansion is relatively small in the chamottes. The mean linear coefficient of thermal expansion up to 1000 °C is from 60 to 70 $\times 10^{-7}$ per °C.

It is well known that the coefficient of thermal conductivity of a material is a function not only of its composition but also of its porosity and of the temperature. The value for normal chamottes with a porosity of about 25% can be taken to be approximately 1 kilocal/m deg hour. The coefficient is temperature dependent, increasing slightly with increasing temperature to a value of about 1,2 at 1000 °C.

Insulating chamottes used especially for lightweight refractory concretes are of very low density and have thermal coefficients well below 1, ranging from 0,4 to 0,2 kilocal/m. deg. hour.

2.3.3. Materials of High Alumina Content

Materials with a high content of Al_2O_3 are made from naturally-occurring high alumina substances, such as cyanite, diaspore or bauxite, or from artificially calcined alumina or corundum. The total of Al_2O_3 and TiO_2 is greater than 45%. The best known of this group are the mullite compounds, consisting chiefly of mullite (A_3S_2) and the corundum chamottes.

2.3.3.1. Chemical and mineralogical composition

Pure mullite, A_3S_2, contains 72% Al_2O_3 and 28% SiO_2. The commercial mullite compounds made by the burning of sillimanites or cyanites vary in their composition between the following limits, depending upon the raw materials used:

SiO_2 14 to 35%   CaO   0,1 to 0,5%
Al_2O_3 62 to 83%   MgO   0,1 to 0,2%
Fe_2O_3 0,5 to 1%

High alumina aggregates made from naturally occurring raw materials consist (see Fig. 18 for the $SiO_2 - Al_2O_3$ system) either of mullite + cristobalite (SiO_2)

or of mullite + corundum. A predominantly mullitic phase is to be expected only in the case of sintered products of standard composition.

Corundum chamottes enriched with smelted corundum are usually manufactured to have standard Al_2O_3 contents of 60, 70, 80 and 90% (denoted by the abbreviations K 60, K 70, K 80, K 90). They consist of suitable quantities of corundum and the proportions of cristobalite, mullite and vitreous material determined by the chamotte and clay fractions.

2.3.3.2. Thermal properties

It can be seen from the phase diagram for $SiO_2 - Al_2O_3$ that in compounds to the left of mullite the eutectic at 1595 °C is the determining factor in the onset of melting. Compounds with a percentage of Al_2O_3 of 72% or greater do not begin to soften until a temperature of 1810 °C is reached. This is valid, however, only for homogeneous mixtures under equilibrium conditions. Corundum chamottes, which may sometimes contain quite large amounts of heterogeneous mixtures of corundum and chamotte grains, connot be judged by this system if their Al_2O_3 content is greater than 72%, since the chamotte phase behaves almost independently, beginning to soften even at 1595 °C.

The cone fusion test points of mullite structures lie in the range 1790 to 1850 °C; for corundum chamottes they lie, depending upon the Al_2O_3 content, between 1790 and 1920 °C. The resistance to pressure at high temperatures of the mullite group of materials is remarkably good. The t_a-values are always above 1600 °C and often reach 1700 °C. The extent of softening under pressure in the corundum chamottes is very much dependent upon the corundum content. In the case of the K 60 compound (60% content of Al_2O_3) a t_a-value of about 1500 °C can normally be expected and a t_e-value of approximately 1700 °C. For an Al_2O_3 content of 80 to 90%, t_a is in the range 1500 to 1600 °C and t_e 1650 to 1700 °C. The allowable service temperature can be assumed to be 1500 to 1600 °C.

One of the great advantages of the mullite materials is their low coefficient of thermal expansion. The mean value between 20 and 1000 °C is approximately 45×10^{-7} per °C. The coefficient of expansion of corundum chamottes varies according to the proportion of corundum present, the coefficient for this constituent being about 80×10^{-7} per °C.

The coefficient of thermal conductivity of mullite at room temperature is approximately 1,4 kilocal/m deg hour and drops slightly with increase of temperature (1,26 at 1100 °C). This property, combined with the low coefficient of thermal expansion, is the basis of good resistance to changes in temperature. Corundum chamottes have a coefficient of thermal conductivity rather larger than that of the normal chamottes; approximately 2 kilocal/m deg hour, dropping to 1,8 at 1000 °C.

2.3.4. Corundum

Corundum is much sought after and used to a considerable extent as an aggregate for the manufacture of fire-resistant concrete, on account of its high melting point, its chemical resistance and its great mechanical strength. In its ground form it also reduces the shrinkage of concrete [216].

Commercial corundum contains between 96 and 99,5% of Al_2O_3. Its smelting temperature when completely pure is 2050 °C. The cone fusion test point of commercial corundum is also of the same order (≈ 1960 °C). Softening under pressure does not commence until well above 1700 °C and therefore cannot be determined exactly.

For the types of corundum used in refractory work the coefficient of thermal expansion varies to a certain extent linearly with temperature: thus the mean coefficient in the range 20 to 1000 °C is approximately 60 to 80 $\times$ 10^{-7} per °C.

The coefficient of thermal conductivity of commercial corundum varies from 2 kilocal/m deg hour at room temperature to 1,8 kilocal/m deg hour at 1000 °C.

2.3.5. Basic Materials with a High Content of Magnesium Oxide or of Calcium and Magnesium Oxide

The materials in this group chiefly consist of sintered or smelted magnesia or sintered dolomite. Not only are they used as aggregates for cement-bonded concretes; they are also the principal basis of the magnesia concretes.

2.3.5.1. Chemical and mineralogical composition

The contents of the most important oxides in sintered magnesia lie in the following ranges:

SiO_2	0,5 to 8%		CaO	2 to 8%
Al_2O_3	0,5 to 2%		MgO	85 to 91%
Fe_2O_3	1 to 8%			

Periclase and magnesioferrite (MF) are the most important mineral phases. Other compounds may be expected, depending upon the quantities of impurities present. Their composition and quantity will depend principally upon the $CaO:SiO_2$ ratio. Amongst the most important are the following: monticellite (CMS), mervinite (C_3MS_2), forsterite (M_2S), spinel (MA), dicalcium silicate (C_2S), dicalcium ferrite (C_2F), tricalcium silicate (C_3S), and brownmillerite (C_4AF).

The following figures may be given for the analysis of sintered dolomite:

SiO_2	1 to 3%		CaO	53 to 60%
Al_2O_3	0,5 to 2%		MgO	34 to 40%
Fe_2O_3	1 to 2%			

The chief compounds are CaO and periclase, but either C_3S, C_3A and C_4AF, or C_3S, C_4AF and C_2F occur as well, depending upon the composition and the $Al_2O_3:Fe_2O_3$ ratio.

2.3.5.2. Thermal properties

The fire-resistance limits of both materials are very high. Magnesia concrete can reach 1980 °C and above. Impurities lower the temperature of softening, since the new phases which may occur melt at relatively low temperatures (for example, CMS at about 1500 °C, C_4AF at about 1450 °C).

Very variable values have been given for the resistance to pressure at high temperature of sintered magnesia. The t_a-values vary between 1450 and 1700 °C and the t_e-values between 1550 and more than 1700 °C. Sintered dolomite structures have t_a-values between 1480 and 1730 °C and t_e-values greater than 1730 °C.

High coefficients of expansion and thermal conductivity are characteristic of minerals with a high MgO content. The mean linear coefficient of thermal expansion for sintered magnesia in the range 20 to 1200 °C is 120 to 140 × 10⁻⁷ per °C and the coefficient of thermal conductivity is between 4 and 5 kilocal/m deg hour at low temperatures and about 2 to 3 kilocal/m deg hour at 1200 °C. The values for sintered dolomite are similar.

2.3.6.   Chrome Ore and Materials of the Chrome Ore-Magnesia Range

Chrome ore is a naturally occurring substance having $FeO \cdot Cr_2O_3$ (chromite) as its principal constituent. Materials made from chrome ore and sintered magnesia, with the former constituting the larger part, are known as chrome ore-magnesia products. Magnesia-chrome ore products, on the other hand, are substances of similar composition, but with a preponderance of sintered magnesia.

2.3.6.1.   Chemical and mineralogical composition

The chemical composition of chrome ore varies over an exceptionally wide range, depending upon the seam from which it is obtained. The following figures may be used as a guide:

Cr_2O_3	28 to 64%	MgO	8 to 26%
FeO	12 to 27%	SiO_2	0 to 5%
Al_2O_3	8 to 31%	CaO	0 to 1%

The analysis of pure iron-chrome spinel must give 32% FeO and 68% Cr_2O_3. Commercial chrome ore is, however, a mixture of chromites of varying composition, magnesiochromite ($MgO \cdot Cr_2O_3$), and hercynite ($FeO \cdot Al_2O_3$). The general formula which can be applied to it is $(FeO, MgO) \cdot (Cr_2O_3, Al_2O_3, Fe_2O_3)$. It may also contain admixtures of magnesium silicate and chlorite.

The high sensitivity of chromite to oxidation should be noted; this leads to the formation of Fe_2O_3 from FeO, resulting in the destruction of the spinel structure. Changes in the atmosphere of the furnace at high temperatures can therefore be dangerous, since the chromite structure is destroyed.

Materials of the chrome ore-magnesia range are mixtures of chrome ore and sintered magnesia and are usually of large grain size. Their mineral phases are formed from those of the separate constituents.

2.3.6.2.   Thermal properties

Pure chromite smelts at 2180 °C, magnesiochromite at 2350 °C, hercynite at 1750 °C. The fire-resistance of chrome-ore structures (bound with a small percentage of magnesia) extends to about 1920 °C. Chrome ore-magnesia products have cone fusion test points higher than 1920 °C.

Values for the resistance to compressive stress at high temperature of chrome ore are known only in combination with sintered magnesia binders. They are relatively low, with $t_a \approx 1550\,°C$ and $t_e \approx 1600\,°C$, but can be attributed to mechanical causes, amongst others, such as rupturing as a result of deformation. The same is true in principle also for products of the chrome ore-magnesia range, for which t_a-values of 1500 to 1600 °C and t_e-values of 1600 to 1730 °C have been given.

The coefficients of thermal expansion for the chrome and aluminium spinels are somewhat lower than that of magnesia. Between 20 and 900 °C they lie in the range 75 to 90 $\times$ 10^{-7} per °C for chromite, 60 to 85 $\times$ 10^{-7} per °C for magnesiochromite, and 80 to 90 $\times$ 10^{-7} per °C for hercynite. Commercial chromite structures have a mean coefficient of thermal expansion of approximately 80 $\times$ 10^{-7} per °C and products of the chrome ore-magnesia range a coefficient of approximately 90 $\times$ 10^{-7} per °C.

Values for the coefficients of thermal conductivity have been stated as follows: for chromite, about 1,2 at 300 °C and 1,45 at 1000 °C, and for chrome magnesia, about 1,85 and 1,25 respectively, these values being in kilocal/m deg hour.

2.3.7. Forsterite

Forsterite masses are materials manufactured from sintered forsterite or naturally-occurring magnesium silicate stone, such as serpentine, olivine or dunite by the addition of caustic or sintered magnesia. Their principal components are forsterite ($2\,MgO \cdot SiO_2$) and periclase.

The chemical composition lies within the following limits:

SiO_2	30 to 38%	Al_2O_3	1 to 8%
MgO	52 to 58%	Fe_2O_3	2 to 8%
CaO	$\approx 1\%$		

Pure forsterite melts at 1890 °C. The equilibrium diagram for MgO—SiO_2 shows, however, that a eutectic exists between MgO and $2\,MgO \cdot SiO_2$ at 1850 °C (63% MgO, 37% SiO_2). Further impurities may have the effect of lowering the softening temperature. The cone fusion test point temperature of commercial forsterite is in general still higher than 1830 °C. Softening under pressure begins at an appreciably lower temperature ($t_a \approx 1600\,°C$, $t_e \approx 1700\,°C$).

The mean linear coefficient of thermal expansion over the range 20 to 1500 °C is approximately 100 $\times$ 10^{-7} per °C. The coefficient of thermal conductivity is about 1,3 kilocal/m deg hour.

The use of naturally occurring olivine, bound into a concrete type material by means of waterglass or magnesia binding agents, is covered by a number of long-standing English patents and has been described by *A. B. Searle* [76].

2.3.8. Silicon Carbide

Commercial silicon carbide approximates to the formula SiC, the presence of impurities being unimportant. It is a hard material having exceptionally good fire-resistant qualities. SiC does not have a melting point, but begins to de-

compose at about 2200 °C. The limit of temperature at which it can be successfully used is also exceptionally high; the t_a-value exceeds 1700 °C. A disadvantage is that it oxidises slowly at high temperatures.

The coefficient of thermal expansion is unusually low, having a value of
only about 23×10^{-7} per °C. In this it differs widely from the cements. A characteristic of silicon carbide is its very high coefficient of thermal conductivity,
the values stated in technical literature on the subject ranging from 10 to 20 kilo-
cal/m deg hour, depending upon the degree of purity.

Pure SiC is normally used for fire-resistant concrete, so that the higher of
the values stated above may be applicable.

2.3.9. Materials possessing a High SiO_2 Content

The most important substances in this group are silica materials and fire-
resistant kieselguhr. As already noted in Section 2.3.1.1, unburnt quartz stone
is unsuitable for use in refractory concrete, on account of the danger of growth.
The only materials of high silicic acid content which can be considered, therefore, are those which have previously been burnt (silica) or which rapidly
undergo transformation when heated (amorphous silicic acid).

2.3.9.1. Silica Materials

As far as the authors are aware, the use of silica as aggregates for cement-
based refractory concretes has not so far been described. The unsuitability of
this material can be attributed to the transformation of crystobalite and quartz
at temperatures of approximately 230 and 575 °C respectively, with the associated large volumetric changes. Aggregates with a high content of silicic acid,
however, have recently begun to be used in the Soviet Union for the manufacture of concrete-type materials using waterglass [102]. They appear to be
suitable also for phosphate-bound materials [84].

The term silica material is understood to mean products obtained by the
burning of quartzite with the addition of lime-water, and consisting principally
of crystobalite and quartz residue. The chemical analysis gives almost pure
SiO_2. According to TGL 4322 [102a], silica stone (bricks) contain approximately 95% SiO_2, 1,5% $Al_2O_3 + TiO_2$, 2% CaO and 1% Fe_2O_3. Their cone
fusion test points are around 1750 °C, t_a-values approximately 1660, and t_e-
values approximately 1680 °C.

The expansion of silica is relatively large, usually being taken to be 1 to
1,4% up to 1000 °C. A steep increase between 200 and 300 °C and a general
flattening of the curve above 600 °C are characteristic. The coefficient of thermal
conductivity at room temperature can be taken to be about 1 kilocal/m deg hour.

2.3.9.2. Kieselguhr

Kieselguhr is also sometimes used for the making of lightweight refractory
concrete. This is a natural, fine-grained, porous mineral mixture, having a high
SiO_2 content. Its main constituent is amorphous silicic acid.

The composition of kieselguhr varies quite considerably, depending upon its place of origin. That which is of good quality will contain very few impurities. The cone fusion test point is about $1670\,°C$ and the coefficient of thermal conductivity is less than $0,1\ kilocal/m\ deg\ hour$. When it is heated above $1000\,°C$, the amorphous SiO_2 crystallises into crystobalite accompanied by a large degree of shrinkage and partial loss of the porous structure.

2.4.   Additives

Additives are understood to include all those added substances that do not fall under the headings of binding agents and aggregates. The principal additives are the very fine-grained, so-called ceramic stabilisers, which are important for high-temperature concretes made with portland cement and are also used in waterglass concretes. Additives also include materials for improving plasticity, accelerating hardening and improvement of the sintering properties when subjected to heat. Various materials which are added for the production of lightweight concrete for high-temperature use (foamed concrete) are also considered under this heading.

2.4.1.   Very Fine-grained Additives ("Ceramic Stabilisers")

The most important additives for high-temperature concretes having a portland cement basis are certain very fine aggregates whose purpose is to bind the free lime. The latter is released first as $Ca(OH)_2$ and above $550\,°C$ in the form of CaO. For concrete-type materials having a waterglass basis, additives of this type improve the bond and strength of the binding agent.

Added materials of this class are distinguished in Soviet literature by the term "very fine additives" in contrast to "normal aggregates" and are dealt with separately in references [8] and [9]. *L. Ludera* [103] has coined the term "ceramic stabilisers" for them, a term which accurately describes their function.

The general requirement for very fine aggregates is a high degree of reactivity to waterglass and lime at high temperatures. This requires on the one hand the correct chemical composition (acid oxide in large quantities) and on the other the fineness necessary to facilitate homogeneous distribution and fast chemical reaction.

Soviet research workers, who have been leaders in studying the effects of ceramic stabilisers, have investigated and found quite a large number of suitable additives. The most important of these are discussed briefly below.

2.4.1.1.   Chamotte

The most important very fine-grained additive, particularly for large scale construction, is chamotte. All the normal and acid types of chamotte, which have been produced by the burning of fire-resistant clays at temperatures higher than $1200\,°C$, are suitable, as well as the various waste products from the production of chamotte. Soviet regulations [9] state that the content of $Al_2O_3 + TiO_2$ shall be greater than 20%, and the SO_3 content (in the case of used acid chamotte) not more than 0,3%; fire resistances should lie between the temperatures of 1610 and 1710 °C. A fineness such that at least 70% passes a sieve of 0,09 mm mesh is specified. It is assumed to be understood that, the finer the chamotte is ground, the higher the strength of the concrete [6, p. 77].

2.4.1.2. Clay

Clay powder or dust fulfills in principle the same functions as chamotte. In addition, its ability to retain water is considered to be advantageous, since it prevents premature drying out. Finally, the plasticising action of clay must be borne in mind.

Fire resistance of the material is essential, a property which is usually achieved with kaolinic clays, particularly, for example, with shale clays [17], [104]. Clay additives have now been used on a considerable scale for a number of years with very good results, both in portland cement concrete and in concrete made from smelted aluminous cement (in the latter case as a micro-filler for improving the plasticity of the wet concrete or as an additive for producing higher strength in the middle temperature range [105]).

2.4.1.3. Chrome ore

Chrome ore, in the form of a very fine grain additive, is used in the U.S.S.R. in the manufacture of high-temperature-resistant concrete with a portland cement basis. Only certain special types are, however, suitable for this purpose. The required content of Cr_2O_3 is at least 45%, while permissible maxima are 8% for SiO_2, 16% for $FeO + Fe_2O_3$, and 1,5% for CaO. It is specified that the grain size should be such that a minimum of 50% and a maximum of 65% should pass a 0,09 mm sieve.

It is stated by *L. A. Cejtlin* [108], however, that from 15 to 25% chrome ore in finely ground form can be added to concretes made with aluminous cement to counteract shrinkage and to increase the strength. In this case the fineness should be less than 0,09 mm.

2.4.1.4. Sintered magnesia

The addition of very finely ground magnesia brick is permitted in the manufacture of high-temperature-resistant portland cement concrete in the Soviet Union [8], provided the material meets the requirements for physical and chemical properties stated in GOST 4689-49. A minimum of 50% must pass a 0,09 mm sieve.

2.4.1.5. Quartz dust and other materials of high silica content

Quartz dust can be added as a ceramic stabiliser without danger, so long as the quantity is kept small. It is necessary, however, to ensure full reaction with the lime. The use of fine quartz sand for portland cement concretes intended for operation at temperatures up to 1200 °C is specified in [7].

Descriptions may also be found in scientific publications of the successful use of kieselguhr, loess, Tripoli powder, and tuff, particularly in the form of pozzolanic portland cements [109], [110]. For the properties of kieselguhr the reader is referred to Section 2.3.9.2. Tripoli powder has a content of more than 80% SiO_2, while tuffs may contain up to 75% SiO_2 depending upon their origin. Loess is an unstratified sediment having the fineness of a dust, the particle size lying between 0,01 and 0,05 mm. Amongst other things it contains 70 to 90% SiO_2 in the form of quartz, together with felspar, mica, clay and limestone fractions. The proportion of clay-like materials may be up to 15%, that of limestone up to 10%. The limiting value of the content of SiO_2 is given in [8] as 70%, and those for Fe_2O_3 and CaO, 8% each. This material is favoured as a micro-filler in portland cement concretes on account of its composition and the nigh reactivity of its constituents [111].

In the manufacture of lightweight refractory concretes, the introduction of previously burnt diatomaceous earth has proved to be advantageous as a highly reactive substance. Finely ground ceramsite has also been used for this purpose [112]. Artificial waste-products, such as silica compounds having about 80% SiO_2 and 15% Al_2O_3, are also suitable as ceramic stabilisers.

2.4.1.6. Fly-ash

Fly-ash is the dust removed from power stations. It is produced in large quantities, but unfortunately varies considerably in its composition. It is, however, of very fine particle size.

The examples quoted in references [6] and [113] of the use of fly-ash as a very fine aggregate, though giving no information about the analyses, probably refer principally to filtered coal-ash. Their compositions lie roughly within the following limits:

SiO_2	40 to 60%	CaO	1 to 10%
Al_2O_3	15 to 30%	MgO	1 to 5%
Fe_2O_3	7 to 30%	SO_3	1 to 9%

Fly-ash should be used, however, only if the sulphate content is very low. It is recommended in [8] that the Al_2O_3 content of fly-ash should be at least 20%, and the sulphate content should not exceed 4%.

2.4.1.7. Slag

Slags from very many sources are stated to be suitable for use as very fine additives. It is essential that they have been ground up, however, otherwise they are not of very much use. The basicity of slags should also meet the requirement

$$\frac{CaO + MgO}{SiO_2 + Al_2O_3} \leqq 1;$$

see ref. [8]. *S. A. Epštejn* [111] has described the use of refractory concretes with slag as a stabiliser.

2.4.2. Plasticisers

Plasticisers are sometimes added to wet concrete for the purpose of reducing the water demand and thus the water/cement ratio. They improve the wetting of solid particles and, as a result of the reduced water demand, increase the workability, strength and impermeability of the concrete. Particularly in the case of portland cement, their use can be treated as normal practice.

A number of commercially available substances are used. Small quantities only are suitable if the best results are to be obtained, varying between 0,1 and 0,2% of the cement content. Higher proportions are usually detrimental, affecting the hardening process. *V. P. Ivanova* [114] has made specific investigations into the effects of various plasticisers on the properties of refractory concretes. Examples of suitable additives are 4% of bentonite, 4 to 6% of fire-resistant clay or mixtures of these two substances, 0,1% of macerated sulphite or 0,005% of a special plasticiser (no further details given). These were for the improvement of workability The strengths of the concretes are admittedly lowered somewhat by such additives, but the loss is compensated by a reduction in shrinkage. Mortars and concretes using plasticisers of this type have been proven in a number of industrial furnaces.

The addition of clay has in general been found to be very effective in improving the plasticity of concrete-type mixes having anhydrous binding agents; for example, phosphate-bound substances.

2.4.3. Hardening Accelerators

Certain additives are sometimes used to accelerate the hardening of portland-type cements, which does not normally start for several hours. These shorten the setting-time considerably and also enable high strengths to be reached quickly. For the purpose of facilitating the early dismantling of shuttering, this method is of importance. Other methods include steam and autoclave treatment. Accelerators also enable concreting work to be carried on at temperatures well below $0\,°C$, since the additives lower the freezing point of water and cause rapid evolution of the heat of hydration.

The best known hardening accelerator is calcium chloride, which, when used in combination with other chlorides, is usually sold under various trade names. The addition of a small percentage (referred to the cement content) can reduce the setting-time to a few minutes, though not always without some loss of final strength and increase of shrinkage. The optimum proportion of calcium chloride additive is between 1 and 3%, depending upon the type of cement; it must always be determined by experiment. The different types of cement behave quite differently in association with calcium chloride [115].

The use of hardening accelerators based upon calcium chloride is permitted only, however, for concretes without steel reinforcement, since solutions of chloride corrode the steel. In addition, it is efficacious only for portland-type cements; it has contrary effects on aluminous cement.

There are numerous other substances that can be used to accelerate the hardening of portland cement under specific conditions, as can be seen from the report by *H. Kühl* [116]. Some of them are, however, difficult to reproduce.

The accelerator usually used for the hardening of waterglass is sodium silico-fluoride, Na_2SiF_6. A number of other substances have similar effects, but are not used to a great extent in the technology of waterglass concrete. Researches carried out in the U.S.S.R. have shown that the optimum quantity of Na_2SiF_6 to be added to waterglass is between 10 and 15% of the latter. Slight variations above and below are of no account, but as soon as the proportion departs appreciably from the range stated the properties of the concrete are adversely affected.

2.4.4. Sintering Agents

The existence of a critical strength region and the resultant temperature limitations on the use of refractory concretes suggest the idea of improving the strength in the middle temperature ranges by means of substances which induce sintering and thus lead to suitable reactions. Research work aimed at achieving this, by *G. Franke* and *F. Kanthak* [107], has shown that even powdered glass is suitable for this purpose. The operating temperature must, however, be at least 850 °C. At lower temperatures the effect of the glass powder on the strength is slight.

2.4.5. Foaming and Gasifying Agents

Foaming agents or gas-producing substances may be used to help in the production of porous or cellular heat-resistant concretes; when added to the mix these agents induce a porous structure.

Soviet sources [117] report the use of glues and rosin in an alkaline solution or in saponine. Researches by *K. Martin* have shown that lauryl pyridine bromide and primary saturated alkyl sulphonate are suitable for use as foaming agents [118]. Other foaming agents, such as spellin, can be used in theory but give less stable foams.

For the production of heat-resistant gas-concretes the well-known procedures of the technology of gasified concretes are followed, involving the addition of aluminium powder to portland cement concrete [119]. This reacts with the lime-alkali mixture, giving off hydrogen which results in the production of a porous structure.

2.5. Water

Water fulfills two important functions in the making of concretes; it makes the dry mixture of aggregates and binding agent workable and also reacts with the binding agent to produce the reactions of hydration and hardening. The water must fulfill certain requirements in order to perform these functions; those concerned with quality will be dealt with only briefly here. The correct proportions of water for good concreting will be discussed in Chapter 3.

It is necessary, in order that the chemistry of the setting process shall not be impaired, that the water used shall not be excessively polluted by organic or inorganic dirt and suspended matter or by acids, bases or other salts. For normal concretes the water should be free of earth, loam and similar impurities.

For refractory concretes, however, this limitation is qualified in that in many concrete mixes clay is deliberately added as a plasticising or stabilising component.

All natural water, mineral water or factory effluents that contain the impurities referred to above in appreciable quantities are therefore unsuitable for use in the making of concrete. Their suitability must be demonstrated by appropriate tests. The strengths of concretes made with water of doubtful quality must not fall more than 15% below that of similar concrete made with drinking water. TGL 11 357 [120] states that any water may be used for portland cement, iron portland cement, or blast furnace cement, provided that the pH value is greater than 4 and that it does not contain more than 35 000 mg/ litre (3,5% by weight) of dissolved salts, independent of the ions which are present. If the pH value is less than 4, the water should be neutralised by the addition of hydrate of lime before use. With these cements, except in the case of prestressed concrete, sea-water may also be used, (the salt content of the Baltic being 7000 mg/litre) (0,7% by weight).* Water which is to be used with smelted aluminous cement, however, must be equivalent in purity to drinking water, since in this case even small quantities of impurities can lead either to a failure to set or to excessively fast setting.

---

*) *Translators' note*: The original edition of this book was written for use in the German Democratic Republic, whose only seaboard is the Baltic Coast. The salinity of the Baltic is considerably less than that of the open seas and oceans.

3. Concrete Mix Design

In the context of the detailed discussion which now follows, the phrase "concrete mix design" signifies the design of concrete mixes for special purpose concretes, with reference to the types of the various constituents and the grading of the aggregates. Now that the basic materials and components of high-temperature concretes have been treated from the standpoint of their chemical and physical properties in Chapter 2, the present chapter discusses the problem from the point of view of granulometry and the relationship of the various constituents to one another. This involves not so much a discussion of the principles of concrete design which are directed towards the achievement of special properties, such as heat-resistance, structural strength in a particular temperature range, or chemical resistance, but rather general, basic problems relating to the effects of the granulometry of the constituents, considered both separately and as a mixture, upon the structural material. These parameters influence a wide range of concrete properties, such as the water demand of the aggregate mixture, the workability of the wet concrete, its ease of compaction, the tendency to segregate, the development of strength and other similar characteristics. Finally, the properties and the behaviour of the hardened concrete itself will be influenced whether in the "raw" condition or during or after thermal treatment, consequent upon the properties of the wet concrete.

In this work the principles of concrete mix design are studied chiefly in the context of cement-bonded concretes. Apart from the fact that the use of high-temperature concretes of this class has so far been much more extensive than that of all other types, most of the experience of concrete mix design has been in this field. At the end of this chapter the use of waterglass concretes is looked at briefly and illustrated by an example.

3.1.   Cements

The cements used in the manufacture of high-temperature concretes are mostly of standard type. The results of researches both in the German Democratic Republic and abroad have shown that high-alumina cements and normal aluminous cements (sintered or smelted), portland cement, iron portland cement and blast-furnace cement are each suitable for certain temperatures and ranges of operating temperature. The use of trass cements and pozzolanic cements and variations of these types has not been common until now and has usually been avoided [6, p. 228]. It has recently been shown, however, that under certain conditions pozzolanic cements may be used for both heat-resistant and normal concretes. The only pozzolanic constituents which are normally considered, however, are those that shrink little or not at all when heated.

There are no special requirements for the granulometry of the cements; the values for fineness laid down in existing standards are valid. Cements which satisfy the requirements of the standards are basically acceptable, though certainly in many cases only the best classes of cement are permitted.

In the Soviet Union cements for high-temperature concretes (aluminous, portland and slag cements) have to satisfy at least the minimum quality standard "400"*. This refers, however, to the Soviet standard GOST 970-61, which, since January 1 1964, has been superseded by GOST 10178-62. The latter reduces the strengths of the cements by approximately 20 to 30%.

In Poland portland cement of "350" and "400" quality is used. In Czechoslovakia aluminous cement, portland cement "350" and "450" and iron portland cement "350" have been tried. There has been extensive use in France of the high aluminous cement "SECAR 250" (75% Al_2O_3) for fire-resistant concretes and of normal commercial aluminous cement (40% Al_2O_3). In West Germany smelted aluminous cements are the most commonly used.

In the German Democratic Republic heat-resistant concretes for both cast-in-situ and precast construction are made from normal commercial aluminous cement to TGL 9738 [58] and from portland cement of quality "350" to TGL 9271 [44]. Standards are laid down for design using both types of cement in TGL Design Memorandum 99-30. In addition, blast-furnace cement has been used successfully for unstressed concrete work, other than the batch production of in-situ and precast concrete [96], [97].

3.2.   Granulometry of the Aggregates

The aggregates used play the largest part in determining the resistance of concrete to high temperatures. The grain size and the grading of the aggregate are of major importance to the quality of refractory concrete.

It is possible by the use of suitable graduations of the aggregate grain size to achieve very dense packing of the aggregate. The grading curves applicable to normal concrete are of only limited use in the present case. A dense, wet concrete prepared to these curves retains its volume on setting and, apart from slight shrinkage, each grain remains in the position it occupied when the concrete was wet. According to the composition of the aggregate, drying and burn-

---

*) *Translators' note:* Qualities relate to strengths in kp/cm².

ing may produce shrinkage or growth of the material. This will result in internal movements and displacements of the grains.

The densest packing which is possible will vary therefore according to the type and make-up of the mix. In general the value of the maximum possible density depends upon the form and grading of the particles. Aggregates of rounded form have a low demand for water. Those that have a needle-like or laminated form, such as are used to a great extent in heat-resistant and fire-resistant concrete, have about twice the surface area of rounded grains. The surface nature of an aggregate has very little effect upon the strength of the concrete [14], but the effect is very noticeable in bending tension. Rough particle surfaces are better for this purpose than smooth. Strengths are still further improved if, in addition to roughness, the surface has a slight tendency to absorb water. This is the case for many of the aggregates used in making heat- and fire-resistant concretes.

3.2.1. Grading of Aggregates

In heat-resistant and fire-resistant concretes the grading of the fine and coarse aggregates must satisfy in principle the same requirements as those specified for normal (other than lightweight) concrete. Particulars are given in Section 3.2.2. The more general aspects are discussed in the following.

3.2.1.1. Questions of maximum aggregate size

In general the maximum permissible size of aggregate will depend upon the minimum size of member to be constructed. Thus *K. D. Nekrasov* [6] gives 40 mm (1,6 in) as the maximum aggregate size for massive concrete construction, and 20 mm (0,8 in) for other civil engineering work. The directions laid down by a certain French firm for refractory concrete construction [121] state that the largest dimension of aggregate should be between one quarter and one fifth of the minimum thickness of member. Irrespective of the minimum thickness, however, the largest piece should not exceed 30 mm (1,2 in).

Table 3 gives the recommendations of *B. Koči* [122] for aggregate sizes in relation to thickness of construction.

Table 3. Aggregate grading related to thickness of construction (Koči)	Thickness of construction [mm]	Grading of aggregate [mm]
	< 20 (0,8 in)	0 to 1
	20 to 50 (0,8 in to 2 in)	0 to 1⎫ 1 to 5⎭
	50 to 100 (2 in to 4 in)	0 to 1⎫ 1 to 5⎬ 5 to 10⎭
	100 to 200 (4 in to 8 in)	0 to 1⎫ 1 to 5⎬ 5 to 20⎭
	> 200 (8 in)	0 to 1⎫ 1 to 10⎬ 10 to 30⎭

Table 4 gives the relationship between grading of aggregate and thickness of construction specified in Poland [123].

Table 4. Aggregate grading related to thickness of construction (Ludera)	Thickness of construction [mm]	Grading of aggregate [mm]
	< 80 (3,2 in)	0 to 2 2 to 5
	80 to 150 (3,2 in to 6 in)	0 to 2 2 to 5 5 to 10
	> 150 (6 in)	0 to 2 2 to 5 5 to 10 10 to 25

The refractory concrete pressings made in the VEB Chamotte Works of Brandis have a maximum aggregate size of 30 mm (1,2 in). In normal dense concrete the maximum dimension should be not greater than one-quarter of the smallest wall thickness and not greater than three-quarters of the distance between reinforcing bars. The following figures for normal concrete are given by *A. Hummel* [124] and *W. Schulze* [14]:

For mass concrete:	passing a sieve of	56 to 125 mm mesh
For reinforced concrete:	passing a sieve of	25 mm mesh
For coarse mortar:	passing a sieve of	5 mm mesh
For fine mortar:	passing a sieve of	2 mm mesh

The reason for the above limits of size of aggregate is that water pockets tend to form on the underside of the larger pieces of aggregate when the concrete is worked, resulting in soft spots in the concrete. By compacting the larger masses of concrete in a rather drier state by means of internal vibrators such points of weakness can, however, be completely avoided.

The following considerations apply to the limiting of maximum aggregate size in heat-resistant and fire-resistant concretes; the amount of working to which the concrete is subjected must be moderate, due to the rough surface of the aggregate and the fairly large amount of water that is sometimes taken up by the aggregate (for example in the case of chamotte and broken brick aggregates). There is therefore a tendency towards the formation of water pockets. In addition, however, and of much greater importance, the aggregates in refractory concretes have another function also, that of initiating the ceramic bonding. Assuming that the correct materials are present for the temperature duty that is required, the reactions will be limited to the surface and boundary layers, should the dimensions of the aggregates be too large. The material will therefore be heterogeneous; because of its variable characteristics it will have internal stresses, leading to the formation of cracks and considerably reducing the quality of the material.

The grading can be ascertained in accordance with TGL 11360 [125] and the shape factor from TGL 9928 [126].

3.2.1.2. Methods of grading aggregates

A. Hummel [124] distinguishes the following types of grading:

(a) Continuous grading, or continuous stages of grain size. The aggregate contains all sizes of particles between the smallest and the largest and the sieve curve is therefore continuous.

(b) Regulation gap-grading. Here certain sizes of particles are missing, or have been deliberately excluded. The sizes omitted are not chosen arbitrarily, but are those where their loss will not affect the compaction of the concrete. The sieve curves are thus discontinuous, in a regular fashion.

(c) Unregulated and arbitrary gap-grading, in which some grain sizes are omitted in an arbitrary way. Their absence may be particularly harmful to the proper compaction of the concrete. The sieve curve is thus discontinuous, in an irregular fashion.

(d) Single-sized mixtures, or aggregates of almost uniform particle size.

The three commonest methods of illustrating the sieve analyses by means of sieve curves are shown in Figs. 19, 20 and 21. The method most widely used is that in which the x-axis is divided linearly, as shown in Fig. 19. The fine aggregate is, however, not clearly described by this system. This objection can be obviated by means of the logarithmic scale. The system shown in Fig. 21, however, depicts the coarse aggregate fraction better, by illustrating the fine fraction logarithmically and the coarse fraction linearly.

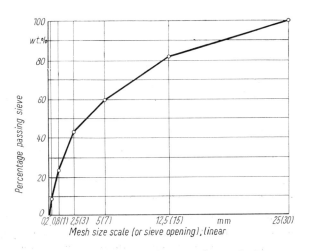

Fig. 19. Grading curve with linear scale on abscissa

The optimum sieve curves for concretes (other than lightweight concretes) are given by the so-called Fuller curves. A distinction must be made, however, between Fuller curves that include the binding agent and those which describe the aggregate alone.

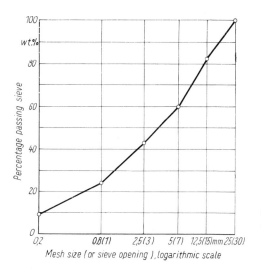

Fig. 20. Grading curve with logarithmic
scale on abscissa

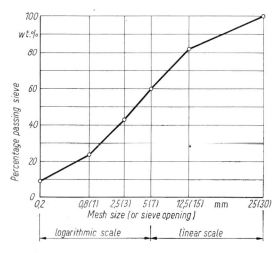

Fig. 21. Grading curve with scale
of abscissa partly logarithmic
and partly linear

3.2.1.3. Limiting grading curves

The so-called limiting grading curves have been developed and are prescribed
in the various standards, because the optimum sieve curves cannot be achieved
by aggregates of equal maximum particle size and aggregates cannot in practice
be made to follow rigidly a particular grading curve.

The grading range for sand of 0 to 5 mm particle size (the fine fractions
for heat-resistant and fire-resistant concretes in this context) are defined in
the standard TGL 0-1045 [127] by the curves A, B and C in Fig. 22. The
corresponding curves for aggregates of 0 to 25 mm (1 in) particle size are similarly defined by the curves D, E and F in Fig. 23.

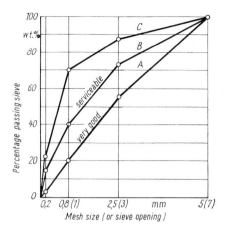

Fig. 22. *Zones for concreting sand to TGL 0—1045*

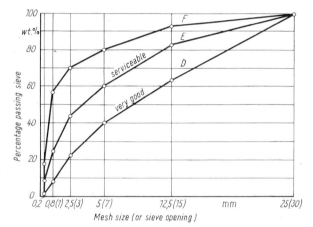

Fig. 23. *Zones for concrete aggregates to TGL 0—1045*

If the sieve curve obtained by a sieve analysis of an aggregate of 0 to 25 mm size lies between the lines D and E — it is allowable that there should be some slight discontinuity in the form of the curve — then this means that the aggregate has a particularly good grading. If the sieve curve lies between E and F, the aggregate exhibits a serviceable grading. Mixtures which lie above line F contain too many fine particles, while those below line D contain an excess of the coarse fraction. The grading for an aggregate in the 0 to 5 mm (0,2 in) range is assessed in exactly the same way.

The figures given in brackets in Fig. 22 and Fig. 23 denote the values which were previously used. The comparison of the old and new values is also given in Table 5.

The influence of the grading upon the strengths of the concretes will be explained by reference to the grading curves D, E and F. The higher the grading curve for an aggregate lies above line D, the lower will be the strength characteristic of the aggregate, and the higher its demand for water for a given concrete stiffness. If the strength characteristic of the aggregate which corresponds to line D is taken to be 100%, then sieve curves which lie above this line will give lower strengths, other factors being equal. This is shown in Fig. 24. It must be remembered in this connection that the reduction in strength due to variations in grading is also influenced by the mixing procedures and the stiffness of the mix [124].

Table 5. Proportions for separate fractions

Grading			
To TGL 10809		Proportions previously used	
Retained by sieve	Passing sieve	Retained by sieve	Passing sieve
—	0,8	—	1
0,8	5	1	7
5	25	7	30
25	56	30	70

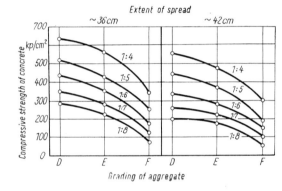

Fig. 24. Relationship between grading of aggregate, compressive strength of concrete, stiffness of mix and mix proportions (after Seidel, from [124])

3.2.1.4. Improvement of grading

The densest packing should be achieved at the least possible expense. The ability of an aggregate to be compacted is thus a function of its grading, shape factor and the quantity of water added and in the case of heat- and fire-resistant concretes it is also partly influenced by the quantity of cement.

If the available aggregates do not meet the grading requirements, it is possible to improve their grading in one of the following ways:
(a) The aggregates are separated out into their separate fractions and then remixed in the proportions necessary to give the required grading. This method is, however, generally only possible in the laboratory.
(b) One or two groups of particles, which are not available in accordance with the grading curve in sufficient quantities, are added.
(c) Two aggregates, the one containing too little of the fine fraction and the other too much, are combined together in suitable proportions.

There are a number of methods of improving the grading of an aggregate; these are discussed in the relevant technical literature, for example the works of *W. Schulze* [14], *A. Hummel* [124], and others.

3.2.1.5. Gap-grading and *F*-value

Gap-grading may be used for concretes designed for high temperatures, as it is for normal concretes. Aggregate mixtures of this type with one or more grading gaps give good density of the resultant concrete and good structural strengths, but result in less workable mixes than continuous gradings, particularly if the fines component is small. The mixtures of aggregate become awkward to handle if the number of grading gaps is too large. Concretes of this type usually have to be mechanically compacted.

A. Hummel has developed the method of calculating *F*-values, based upon the concept of the grading factor introduced by *Abrams*. This method makes it possible to judge whether discontinuous gradings will give the same concrete strengths, other factors being equal, as continuous gradings.

If a logarithmic scale is chosen for the axis which indicates aggregate size, then all the sieve curves that show equal surface areas above the basic particle-size curve will have the same possible strength properties as concretes. The *F*-value may be found either graphically or by calculation. Table 6 shows the separate aggregate-size factors to be used in obtaining the *F*-value by calculation. The formula is:

$$F = 100 \cdot \log 10\, d$$

where *d* is the diameter (in mm) of the average ideal grain of aggregate.

Table 6. Separate aggregate-size factors for calculating the F-value, after Hummel	Aggregate size [mm]	Aggregate-size factor
	12,5 to 25	233
	5 to 12,5	201
	2,5 to 5	166
	0,8 to 2,5	124
	0,2 to 0,8	65
	0,1 to 0,2	15

The percentages of aggregates retained between sieves of successive fineness are multiplied by the aggregate-size factors from Table 6, the products are added up and the resultant sum divided by 100.

The following is a worked example:

Aggregate size [mm]	Mass [%]	Aggregate-size Factor		Products
12,5 to 25	18	·	233 =	4194
5 to 12,5	22	·	201 =	4386
2,5 to 5	17	·	166 =	2822
0,8 to 2,5	19	·	124 =	2356
0,2 to 0,8	15	·	65 =	975
0,1 to 0,2	9	·	15 =	135
				14868

This gives $F = \dfrac{14\,868}{100} \approx 149$ cm². This value is then compared with and assessed against the F-values of the limiting grading curves. The F-values of grading curves A to F are given in Table 7 [124, p. 72].

Table 7. F-values of grading curves A to F, after Hummel

Grading curve	F-value
A	130
B	105
C	78
D	184
E	149
F	105

The limits obtained by the F-value calculation are, however, already obvious, since the aggregate in the range of particle size 0,8 to 2,5, having an aggregate-size factor of 124 cm² (see example above) lies between the lines indicating good and serviceable mixtures. A. Hummel has shown from tests that, for equal values of fines, equal values of strength can be obtained in practice only if the aggregate consists of at least three grades of material; one of these should be a fine grading, which will be decisive in determining the aggregate surface area and also the workability of the concrete.

3.2.2. Special Requirements of the Composition of Aggregates for Heat-resistant and Fire-resistant Concretes

The aggregate composition of normal (dense) concrete has been treated so fully in the preceding section because, in the words of the Soviet authority K. D. Nekrasov, the requirements in his own country, one of the foremost in the development of heat-resistant and fire-resistant concretes, are that "the fine and coarse aggregates to be used in heat-resistant concrete must conform to the same requirements as the aggregates for normal concrete, so far as grading is concerned" [6, p. 223].

Special requirements are, however, laid down in some other countries, which differ from those above; something will therefore be said on this subject below.

As already noted, a properly designed mixture of aggregates is required, in order to ensure a dense, well-consolidated concrete and to achieve high structural strength. This is especially true for heat- and fire-resistant concretes, since at about 400 °C the hydraulic bond decreases. Ceramic bonding begins, as is well known, at about 1000 °C. As a result of this change in the nature of the bond in the concrete, all cement-based high-temperature concretes pass through a region of minimum compressive strength. It is evident that in this critical region the binding agent plays a less important role. The aggregates in refractory concrete therefore have an additional function to perform, which does not need to be considered in the case of normal concrete. As noted previously when discussing the reaction of aggregates with one another and the question of maximum density of packing, a higher proportion of fine aggregate below 0,2 mm is required for this purpose. It will be recalled that the aggregate has to act as a micro-filling agent or ceramic stabiliser, improving the strength of the concrete in this critical region.

For these reasons the aggregate proportions established for gravel concretes from the Fuller curve and from the grading curves given in the standard TGL 0-1045 are only partly relevant to the design of mixes for heat- and fire-resistant concretes.

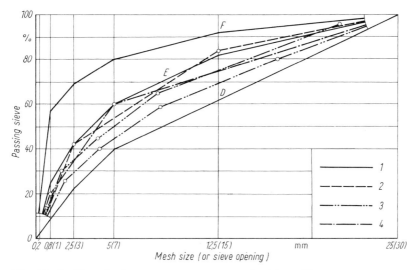

Fig. 25. Grading of refractory concretes compared with grading curves to TGL 0-1045

1 Grading Curves D, E, F TGL 0—1045; 2 after Schmeisser; 3 after J. Arnould; 4 VEB Chamotte Works, Brandis

Fig. 25 shows a comparison between the aggregate proportions for a number of refractory concretes obtained from examples in the technical literature and from the sieve curves given in TGL 0-1045.

3.2.2.1. Particle-size distribution from the Litzov curve

The aggregate proportions used at the VEB Silica and Chamotte Factory at Rietschen for making precast units of refractory concrete, are stated by *E. Kuntzsch* and *G. Rabe* [128] to follow a curve known as the Litzov Curve.

K. Litzov discovered this parabolic curve and, with it, the densest packing obtainable with chamotte aggregates comprising approximately equal quantities of aggregate from three separate sizes. An attempt was made to extend the model concept of the packing of spherical balls to the problem of determining experimentally the best grading for an extremely irregular aggregate of this type. The values

$$\alpha = 2 \text{ to } 5 \text{ mm}; \qquad \beta = 0,25 \text{ to } 2 \text{ mm}; \qquad \gamma < 0,25 \text{ mm}$$

gave a distinct maximum packing density of the material, and of the fired refractory blocks which could be made from it by the addition of 20% of binding clay, using the proportions

$$\left(\frac{\alpha}{2} + \frac{\beta}{2}\right) \approx 70\%; \qquad \gamma \approx 30\% \quad [2;\ p.\ 82].$$

The Fuller curve for normal concrete aggregates and the Litzov curve for chamotte are compared in Fig. 26 [129].

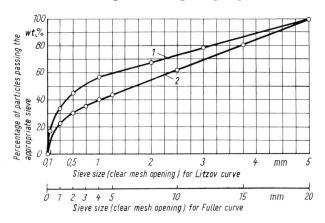

Fig. 26. *Fuller curve for normal concrete aggregates and the Litzov curve for chamottes (from Ref. [129])*
1 Litzov curve for chamotte; *2* Fuller curve for concrete aggregates

V. *Skola* (quoted in [129]) found that the curve of densest packing for chamotte approached more closely to the Fuller curve. This also showed the curve of density of chamotte grains and that of the density of the products manufactured from the chamotte to be generally parallel. In this connection it should be mentioned that for these materials, as for normal aggregates, the densest packing was obtained by the exclusive use of the coarse and fine aggregates, omitting those of middle size, in other words by the use of gap-grading. The higher proportion of fine aggregate indicated by the Litzov curve is confirmed by P. P. *Budnikov* [10] and other Soviet research workers. The finest dust-like grains of chamotte, having a grain-size of 0,2 to 0,1 mm, are considered to be the deciding factor. Increases in the proportion of this fine fraction result in increase of strength, of uniformity and of impermeability in the resultant product, even if the maximum limit of aggregate size is increased to 4 or 5 mm. Admittedly the resistance to temperature change is reduced, but not to a limiting extent, since the breakdown of the structure and the reduction of strength also have a harmful effect upon the stability under variable temperature. The negative influence on the resistance to varying temperatures hardly needs to be considered in a heat-resistant concrete, however, since *Budnikov* assumes a much higher proportion of binding clay as the raw clay component in the form of a micro-filler or ceramic stabiliser in the case of high-temperature-resistant concretes.

3.2.2.2. The Bolomey equation

According to *L. Ludera* [103] the Bolomey equation

$$Y = A + (100 - A) \cdot \left(\frac{d}{D}\right)^{1/2}$$

is recommended for the grading of aggregates. In this equation,

Y = percentage passing sieve,

d = diameter of the fraction of aggregate [mm],

D = diameter of largest grain of aggregate [mm],

A = constant for refractory concretes, varying between 14 and 30 (generally about 25).

For normal structural concretes this constant lies between 10 and 14.

J. Bolomey [130] uses this equation to identify the best grading of the mixture of aggregate and cement, meaning in this context that composition which gives a concrete having good plasticity and high impermeability with such a proportion of fine aggregate that the water demand is a minimum.

With regard to the constant A *Bolomey* shows in an an earlier publication [131] the advantage of modifying the Fuller parabola by the introduction of a constant. This takes the cement and the fines of diameter less than 0,5 mm into account and its value will depend upon the type of materials and the plasticity that is required. In terms of practical manufacture of concrete, therefore, this involves determining the ideal granulometric composition by the *Bolomey* equation, using a value of A which is suitable for the desired plasticity. It is possible to depart appreciably from the theoretical composition of the aggregates without seriously affecting the quality of the concrete, so long as the correct proportion of fine material less than 0,5 mm diameter, the ratio of fine and coarse aggregates, and the allowable maximum grain size are all retained.

3.2.2.3. Aggregate composition for concrete made with broken brick

Figs. 27 and 28 show the grading curves for broken brick concrete of a dense composition, as specified in existing standards. The need for these curves arose

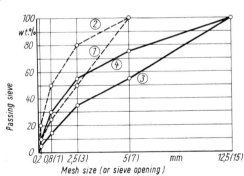

Fig. 27. Grading limits of aggregates for concrete of dense internal structure

1 and *2* for broken brick, size 0—7 mm;
3 and *4* for broken brick, size 0—15 mm

from the fact that the grading envelopes specified in TGL 0-1045 proved to be rather too wide for broken brick concrete. The lower limits of the latter would give a concrete that is rather too difficult to work, being incapable of complete compaction even under heavy tamping, if normal proportions were to be used. There would also be the attendant risk of crushing the aggregate. The upper limits are undesirably high, because the brick dust in the fine fraction would require too much cement.

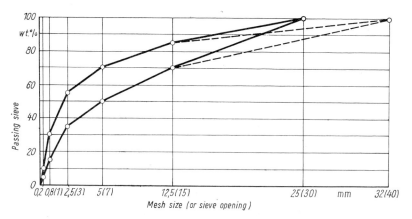

Fig. 28. Grading limits of aggregates for concrete of dense internal structure
Broken brick 0—30 mm and 0—40 mm

3.2.3. Crushing and Sieving of Aggregates

Crushing has an important effect upon the shape of the aggregate and the way that it packs together and consequently also upon the physical properties of the resultant refractory concrete. The subject of crushing is not discussed in this work; reference should be made to the relevant technical literature for a general treatment of the subject (see, for example, *C. Mittag* [132] and *H. Kirchberg* [133]). Here only a few procedures are considered which will serve as a guide and reference point for those cases in which the concrete has to be mixed on site and placed in situ. Precast and ready-mixed concrete are not discussed.

The choice of aggregate for grinding should always be directed towards achieving the optimum in the final product, with regard to certain specific properties, by means of attention to grading [1]. The amount of crushing and the hardness and toughness of the material to be crushed are relevant in this context. It is advisable to employ tough grinders for hard aggregates and hard grinders for soft aggregates. Table 8 gives a summary of the most important types of crushing plant or machinery, related to type of crushing and the raw material to be used.

Practice has shown that multi-stage crushing is preferable; for instance, there may be a preliminary stage of coarse breaking down, followed by crushing and then grinding. The operation proves more efficient if the finest fraction is

Table 8. Crushing and types of crushing plant (from [1])
(D) = Dry crushing: (W) = Wet crushing

Method of crushing and fineness of crushed material	Property of crushed material		
	Hard	Medium hard	Soft
Coarse fraction > 50 mm (2 in)	Jaw crusher (D) Rotary crusher (D) Symons crusher (D)	Jaw crusher (D) Rotary crusher (D) Roller crusher (D) Double-hammer crusher (D)	Hammer crusher (D) Saw (D)
Fine fraction 5 to 50 mm (0,2 to 2 in)	Jaw crusher (D) Rotary crusher (D) Roller crusher (D) Hammer crusher (D) Impact breaker (D)	Jaw crusher (D) Roller crusher (D) Steam-breaker (D) Worm (D) Hammer crusher (D) Impact breaker (D) Steam breaking (D)	Cog crusher (D) Worm (D)
Grit, chippings 0,5 to 5 mm (0,02 to 0,2 in)	Roller mill (D, W) Ball mill (D, W) Pan grinder (D, W)	Roller mill (D, W) Ball mill (D, W) Hammer mill (D) Impact breaker (D) Impact plate mill (D, W) Vibrating mill (D) Disintegrator (D) Jet impact breaker (D) Steam breaking (D)	Hammer mill (D) Impact plate mill (D, W) Sieve mill (D) Cog mill (D, W) Crushing mill (D, W) Cylinder mill (D) Steam breaking (D)

removed from each stage. This will result in a modification to the proportions. For example, *K. Konopicky* [1] has shown that sieving will produce an aggregate in the 0 to 2 mm range which is appreciably coarser than that which results from grinding down to a maximum of 2 mm.

The moisture content of the raw material will have a profound effect upon the grinding and sieving processes, due to the tendency towards internal adhesion and sticking to the sieves. Ball mills, which are often used, are particularly preferred for the finest stage of grinding. The coarse material is usually sieved in rotating screens, the finer ones in one or more layers of vibrating sieves.

3.3. Very Fine Aggregates

As already seen in the section on basic materials and constituents, the very fine aggregates, of a size less than 0,2 mm, have a special function in concretes for use at high temperatures; this aspect is discussed in more detail in the section on the properties of high temperature concretes. Their principal function is of a chemical nature, hence the description "ceramic stabilisers", which is also used in the terminology of refractory concrete. For certain structural concretes fine fractions of this type are of importance in the obtaining of special characteristics. From this point of view they are described as fillers or micro-fillers (TGL 10 809 [133a])

This section is concerned with the granulometric, not the chemical, functions of the very fine aggregates, so the detailed aspects of particle size and pro-portioning are now discussed.

The studies carried out by *H. Salmang* [129] on the subject of the densest packing that can be achieved with chamotte, using only the coarse and fine fractions and rejecting the middle ranges, have shown that the addition of 12% of binding clay does not lead to a loosening effect. *K. Konopicky* [1] gives the following general rule for obtaining the densest packing; the addition of 10 to 15% of binding clay will not cause loosening of the structure, because the clay will chiefly occupy the voids in the aggregate. Under normal circum-stances the porosity of aggregates of this type can reach the value of 16% in practice. It should, however, be possible in theory to achieve a porosity of about 6% by the use of a number of grain sizes [1].

The stabilising effect of the very fine fractions in high temperature concretes can be considered from a similar viewpoint, since the same results will be achieved, at least in physical terms. An additional beneficial effect of clay and materials which contain clay is that during the hardening process they retain the water that is required to facilitate hardening.

For most ceramic stabilisers the optimum particle size varies considerably. *L. Ludera* [103] obtained the best results with the 0 to 0,1 mm fraction, while *K. D. Nekrasov* [6] quotes, for example, a finely ground quartz of less than 0,06 mm as giving the densest structure.

The quantity of the very fine fraction that will be required varies considerably according to its type; it ranges between 30 and 100% of the mass of the cement [6]. *G. Franke* and *F. Kanthak* [107] have obtained good results by the addition of raw clay. *H. Gibbels* [55] has more recently achieved good results with the use of fire-resistant clay of less than 0,75 mm size; the clay was added in the proportion of from 5 to 20% of the total mix for a heat-resistant concrete based upon blast-furnace cement.

Chamotte powder, such as the dust from rotating kilns, gives good results as a ceramic stabiliser, if it is used in the particle-size range 0,06 to 0,2 mm. The proportion added can range up to that of the cement content. The result is to reduce shrinkage, to lessen the disparity between the aggregate and the cement matrix and to increase the compressive strength of the concrete to a certain extent. It is specified in the Soviet Union that the grain size of very fine chamotte aggregate or of aggregate composed of semi-acidic material shall be such that not more than 70% passes a 4900-mesh sieve [9]. The same stand-ard specifies that not less than 50% and not more than 65% of chromite used for this purpose shall pass a 4900-sieve and that the minimum proportion for magnesia shall be 50%. As in the case of chamotte and other fillers, the exact quantity required will depend upon the thermal characteristics needed in the final concrete. It is therefore not possible to give universally applicable rules.

3.4.   Composition of Concrete

A scrutiny of the literature on the subject of concretes for use at high temperatures leads to the conclusion that there is no uniformity about the proportions or other information given on the mixtures to be used. This would appear, however, to be indispensable for obtaining the correct composition. Before discussing the composition of concrete and giving examples, therefore, something must first be said on the subject of mix proportions as the basis of quantitative concrete design.

3.4.1.   Questions of Mix and Proportions

The large-scale development of concretes for use at high temperatures has its origin in the refractory industry. In this industry it is customary to express the build-up of a mix in terms of percentages. The greater part of the literature on concrete follows the same procedure; that is, the proportions of the solid constituents are usually expressed in percentages. Another method of description, found in earlier writings on normal concrete, is to state a "mix proportion" of cement to aggregate. Both types of information are inexact for concretes, since they do not take into account the water content and the subsequent strength is very much dependent upon the amount of water added. For heat-resistant and fire-resistant concretes it is therefore more correct, as indeed it is for all other types of concrete, to express the composition of a mix as a proportion of the following form:

$$\text{Cement:Aggregate:Water} = z:k:w.$$

Alternatively, if the cement content is taken to be unity, this form may be used:

$$1:\frac{k}{z}:\frac{w}{z}.$$

When ceramic stabilisers are used they are included in the aggregate component.

It has become common practice in the case of normal concretes to describe the proportions of the mix in terms of mass, since volumetric proportioning can lead to ambiguities. The description "loosely packed" is not a clear definition of measurement. This is particularly true for aggregates, but to a certain extent applies also in the case of cements. Measuring the proportions by mass can also give rise to variations, but they are of a much smaller order, in the case of normal concretes, than those which occur when the proportions are measured by volume. When proportioning by mass, fluctuations occur as a result of variations in the specific gravity of the aggregate, for which the value is normally about 2,6, and of changes in the moisture content, which is usually not greater than $\pm 3\%$ of the aggregate mass.

In high-temperature concretes, there are many aspects which have to be considered when designing the mix. The main reason for this is the considerable range in the properties of the various aggregates. For particularly dense aggregates, which take up a small amount of water, the same assumptions may be made as for normal concretes in which the proportions are measured by mass. If two or more aggregates of similar structure but having different specific gravities are used in a mix, these differences must be taken into account in the design when calculating the weights of the various constituents required per unit volume of concrete. Also, care must be taken, when weighing the different fractions of the aggregate, that the quantity of each fraction is volumetrically correct, since measures of volume are used when classifying an aggregate according to the amounts passing the various sizes of sieve, whereas the results of sieve tests are given in terms of mass. If all the aggregates in the range used have the same density, then the mass percentages may be used. If, however, the densities of the size groupes vary, then the volumetric percentage of each group must be multiplied by the specific gravity of that group.

The design of high-temperature concretes with porous aggregates, of which broken brick is an example, must use different assumptions from those employed for dense aggregates. When the concrete is being worked the inner pores of each grain are largely unaffected, only the surface pores being penetrated by the cement paste and the very fine (powdered) aggregate. Therefore the density, not the specific gravity, should be used in calculations relating to aggregates of this type.

There are a number of methods of determining the densities of aggregates. A useful method, which also enables the water content of the aggregate to be obtained, is that described in DAMW-N 25-260 [134]. All the well-known methods enable a more or less accurate determination of density to be made for the coarse fraction. Since, however, the densities of the finer fractions of porous aggregates can be either higher or lower (this is a well-known fact), the value of density ascertained for the coarser fractions is not valid for the finer. In practice it is almost as impossible to establish the density of fine-grained porous aggregates as it is to ascertain their absorption of water. For example, if the water absorption of a broken brick aggregate in the 3 to 7 mm (0,12 to 0,28 in) range is established and the value is then used in the design of a mix containing a fine fraction of sand in the 0 to 3 mm range also obtained from broken brick, the resulting mix will be too dry to be workable. Additional water will therefore be necessary for the fine fraction. It is impossible to determine how much water is really taken up by the soaking of the separate grains of the fine aggregate, to the point of saturation, and how much adheres between the grains, resulting in their forming a tenacious paste. The indeterminability of this "excess water" means that it is not possible to design an exact mix in one operation. Concretes which are to contain porous fine aggregates and are to have specific properties, must therefore be designed for workability upon the basis of experiment. This is particularly true for all heat-resistant concretes that contain ceramic stabilisiers. Only experiment can decide whether the percentage of this material is small enough to be ignored. These differences can be reduced by classing the ceramic stabiliser as part of the cement content of the mix; this can be confirmed by reference to the details of portland cement-based, heat-resistant concretes described by L. Ludera [123].

3.4.2. Design Examples

Examples are first given of the mix design of a heat-resistant concrete, using first a dense and second a porous aggregate. In both cases a dense concrete, having a 28-day cold strength of 225 kp/cm² (3200 psi), is required, for an unlined reinforced concrete chimney.The maximum operating temperature is 500 °C. The dense aggregates are to be crushed basalt sand (0 to 3 mm (0,12 in)) and broken basalt (7 to 15 mm (0,28 to 0,6 in)), while the porous aggregates are fine broken brick (0 to 7 mm (0,28 in)) and broken brick (7 to 30 mm (0,28 to 1,2 in)). The binding agent is portland cement, class 350. Because of the relatively low working temperature, a specific ceramic stabiliser is not used. Finally, an example is given of mix design for a heat-resistant concrete on the basis used in the Soviet Union [293].

3.4.2.1. Basalt concrete

The design is based upon the principles adopted by *S. Röbert* [135] for a stable mortar and concrete. These state that a mortar and a concrete which achieve the calculated, predicted, wet concrete compaction when a particular compacting equipment is used, are defined as stable in relation to that compacting equipment. If trial mixes are prepared, having a particular water/cement ratio and a number of different mix proportions and the compactions of these different mixes are then ascertained by experiment, the resultant curve may be plotted. This curve will then intersect the predicted curve of compaction in a certain point. This point then represents the mix which will give the stable mortar or concrete.

The use of a vibrating table (3000 vibrations per minute; vibrating time = 1 minute) results in a mix ratio of 1:1,7 (cement: fine or coarse aggregate of 0 to 3 mm (0,12 in)). The water/cement ratio of 0,6 was taken from the normal relationship between water/cement ratio and strength for ordinary concrete. It was found that coarse aggregate could be added to the mortar up to a maximum proportion of 3,1 without loss of strength or workability. Gap-grading was used for the coarse aggregate. The feasibility and advantages of doing this have already been discussed in Section 3.2.1.1. The resultant overall mix proportions measured by weight were then as follows:

cement:aggregate:water = 1:4,8:0,6 = 1:1,7 (0 to 3 mm)
+ 3,1 (7 to 15 mm):0,6.

The quantities required for one cubic metre of concrete were as follows:
405 kg cement, class 350 (702 lb/yd³),
689 kg crushed basalt sand, 0 to 3 mm (1197 lb/yd³),
1255 kg broken basalt, 7 to 15 mm (2175 lb/yd³),
242 kg water (421 lb/yd³).

The values ascertained for compressive strength were as follows:
after   7 days: 195 kp/cm² (2770 psi),
after 28 days: 272 kp/cm² (3870 psi)
(obtained from tests on 10 cm (4 in) cubes, allowing a reduction of 15% to correlate the values to those for 20 cm (8 in) cubes).

3.4.2.2. Broken brick concrete

Theoretical design of a mortar using crushed brick dust is not possible, because, as has been seen, the density and the water absorption of this type of material are not well known. The correct proportions of the mix must therefore be found empirically. The criteria are good workability without segregation and adequate compressive strength. A value of the water/cement ratio therefore has to be found by experiment which, in spite of the absorption of water by the aggregates, gives good workability and easy compaction of the concrete after allowing it to stand for 30 minutes. Sieve analyses show that the grading curve of the fraction below 15 mm (0,6 in) lies partly in the region of particularly good grading, while that for the 15 to 30 mm (0,6 to 1,2 in) fraction lies close to the grading curve *E*. In this case the aggregates may therefore be used without particular selection. The proportions of the mix, by weight, were found to be:

$$\text{cement:aggregate:water} = 1:4:1 = 1:2 \ (0 \text{ to } 7 \text{ mm})$$
$$+ \ 2 \ (7 \text{ to } 30 \text{ mm}):1.$$

The quantities required for 1 cubic metre were found to be:

341 kg cement, class 350 (591 lb/yd^3),
682 kg crushed brick dust, 0 to 7 mm (1182 lb/yd^3),
682 kg broken brick, 7 to 30 mm (1182 lb/yd^3),
341 kg water (591 lb/yd^3).

The strength was found to have the values:

145 kp/cm^2 (2070 psi) at 7 days,
241 kp/cm^2 (3430 psi) at 28 days

(obtained from tests on 10 cm (4 in) cubes allowing for a 15% reduction when compared with cubes of 20 cm (8 in)).

3.4.2.3. Heat-resistant concrete with chamotte and a micro-filling agent

A concrete is required which is to have B 200 quality (2844 psi) and is intended to be used up to 1100 °C. It is to be based upon portland cement. The slump of the wet concrete should be 2 cm (0,8 in).

A chamotte with a fire-resistance value of 1610 °C and a loose density in the broken state of 920 kg/m^3 (591 lb/ft^3) is chosen for aggregate and also micro-filler. The ratio of the chamotte-dust fraction (micro-filler) to the cement content should be 1 and the ratio of the chamotte dust to the remainder of the aggregate should be 0,4. The proportions required for weigh-batching are established for three different mixes, having cement contents of 250, 300 and 350 kg per cubic metre of concrete (433, 520 and 606 lb/yd^3).

When making the three trial mixes the values of the water contents are determined so that the required slump of 2 cm (0,8 in) is achieved. A minimum of three 10-cm (4 in) cubes is prepared from each mix and the wet concrete densities are determined. The mix proportions can then finally be re-calculated.

Let the values obtained for the cement contents per cubic metre for the three trial mixes in the foregoing example be 247, 293 and 344 kg (427, 507, and 596 lb/yd^3) respectively and the compressive cube strengths obtained from tests be 130, 180 and 245 kp/cm^2 (1850, 2560, and 3480 psi). The values of compressive strength are then plotted against cement content (see Fig. 28a). A straight line is then drawn through these points. The cement content required for a concrete strength of 200 kg/cm^2 (2844 psi) is then read off. In the example given the cement required is 312 kg (540 lb/yd³). The required water content is then determined by interpolation.

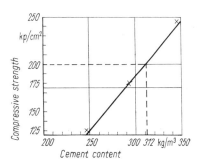

Fig. 28a. Relationship between the compressive strength and cement content for heat-resistant concrete (design example)
concrete in unfired state

The quantities required for 1 cubic metre of concrete are found to be:

      312 kg portland cement (540 lb/yd³),
      312 kg chamotte dust (540 lb/yd³),
      420 kg chamotte sand (730 lb/yd³),
      630 kg broken chamotte (1090 lb/yd³),
      350 kg water (606 lb/yd³).

3.4.3. Normal Compositions of Concrete for High Temperatures in the German Democratic Republic and in other Countries

As mentioned previously the composition of a concrete is defined by stating the proportions of all the constituents of the mix. It is only in this way that a full comparison can be made of different mixes with respect to their usefulness and suitability. Unfortunately, however, it is not always possible to obtain all the proportions of mixes quoted in technical literature, because the information is sometimes incomplete. The compositions of concretes commonly used for high temperatures in the German Democratic Republic and in other countries may, however, be quoted, in order to give the reader a general summary and some useful figures.

3.4.3.1. Heat-resistant concrete for precast refractory elements made by the VEB Silica and Chamotte Works, Rietschen

The composition of all aggregates is obtained from the Litzov curve. The following composition is chosen, using smelted aluminous cement to the standard TGL 9738, for working temperatures up to 1300 °C:

      50 kg (110 lb) smelted aluminous cement,
      110 kg (243 lb) raw chamotte, 0 to 3 mm (0,12 in),
      110 kg (243 lb) raw chamotte, 3 to 8 mm (0,12 to 0,32 in)
      approximate water/cement ratio 0,6.

The following composition is chosen, using portland cement class 350 to the standard TGL 9271, for working temperatures up to 1100 °C:

    50 kg (110 lb) portland cement, class 350,
    57 kg (126 lb) raw chamotte, 0 to 3 mm (0,12 in),
    90 kg (198 lb) raw chamotte, 3 to 8 mm (0,12 to 0,32 in),
    60 kg (132 lb) blast furnace pumice 0 to 3 mm (0,12 in),
    13 kg (29 lb) clay, 0 to 1 mm (0,04 in),
    approximate water/cement ratio 0,7.

3.4.3.2. Mix for heat-resistant concrete supplied by the VEB Chamotte Works, Brandis

The mixes supplied by the above company are composed of chamotte and either aluminous cement or portland cement. The ratio of raw chamotte to cement is usually 80/20 (per cent by weight). In accordance with the standard TGL 99-30 [5] two types of raw chamotte are used, which must have the following cone fusion test points:

    Raw chamotte type I, at least 1730 °C.
    Raw chamotte type II, at least 1680 °C.

The mixes contain different fractions, lying in the following ranges for mixes of different types: 0 to 1 mm (0,04 in), 0 to 3 mm (0,12 in), 0 to 5 mm (0,2 in), 0 to 15 mm (0,6 in), 0 to 30 mm (1,2 in). In addition to chamotte, from 7 to 10% of clay up to 1 mm particle size is added to the portland cement mixes. The water quantities are given as 22 litre/100 kg (water ratio 0,22) for concrete pressings and 24 litre/100 kg (water ratio 0,24) for casting concrete. The maximum operating temperature for concrete having a smelted aluminous cement basis is 1250 °C.

3.4.3.3. Composition of heat-resistant concretes in Poland

Using exclusively portland cement, class 350, the following concrete mixes are made in Poland for maximum operating temperatures of 1150 °C. The rules relating to thick ness of structural components are those already given in Table 4. (Quantities in weights per cent) [123].

Concrete composition for a wall thickness of up to 80 mm (3,2 in):

    Cement with micro-filler material   27%,
    Broken chamotte 0 to 2 mm (0,08 in)   36%,
    Broken chamotte 2 to 5 mm (0,2 in)   37%.

Concrete composition for a wall thickness of up to 150 mm (6 in):

    Cement with micro-filler material   27%,
    Broken chamotte 0 to 2 mm (0,08 in)   19%,
    Broken chamotte 2 to 5 mm (0,2 in)   22%,
    Broken chamotte 5 to 10 mm (0,4 in)   32%.

Concrete composition for a wall thickness greater than 150 mm (6 in):

Cement with micro-filler material	27%,
Broken chamotte 0 to 2 mm (0,08 in)	15%,
Broken chamotte 2 to 5 mm (0,2 in)	17%,
Broken chamotte 5 to 10 mm (0,4 in)	24%,
Broken chamotte 10 to 25 mm (1 in)	17%.

The water content is stated to be 12%. Chamotte dust of particle size 0 to 0,6 mm is used as micro-filler. The quantity is 7%, so that there is 20% of cement in each mix.

The chamotte aggregates used should have a cone fusion test value of not less than 1700 °C. The grading curves correspond approximately to the Litzov curve and lie close to the curves for aggregates for normal dense concretes.

3.4.3.4. Composition of a heat-resistant concrete using andesite as aggregate

A concrete which can be used at temperatures up to 700 °C has been designed for the East Slovakian Steelworks and has already been proved successfully on a full commercial scale [294]. The following quantities are recommended for making 1 cubic metre of wet concrete:

400 kg portland cement, class 350, with 10% of active pozzolana
(695 lb/yd³),

1500 kg andesite rock (2600 lb/yd³),

proportioned thus:	0 to 3 mm (0,12 in)	22%,
	3 to 8 mm (0,32 in)	28%,
	8 to 15 mm (0,6 in)	30%,
	15 to 25 mm (1 in)	20%,

100 kg of fly-ash (ground) (174 lb/yd³),

240 kg of water (417 lb/yd³).

The 28-day strength of this concrete before being put into service is about 250 kp/cm² (3560 psi) the strength after operating at a temperature of 600 °C is approximately 170 (2420 psi) and after reaching 700 °C, 135 kp/cm² (1920 psi).

3.4.3.5. Composition of a heat-resistant concrete with copper slag as aggregate

The following composition permits an operating temperature of 800 °C, if direct contact with slag and alkalis can be prevented. Provided that this condition is observed the concrete can be used even at high stresses [295]. The constituents are as follows:

Binding agent: portland cement, class 375.

Ceramic stabiliser: rotating kiln dust from the manufacture of raw chamotte, up to 0,2 mm (0,008 in).

Aggregate: broken raw copper slag.

The grading according to the Litzov curve (in weight per cent) is:

0 to 0,2 mm	(0,008 in)	16%,
0,2 to 1 (0,8) mm	(0,04 in)	32%,
1 to 3 (2,5) mm	(0,12 in)	19%,
3 to 7 (5) mm	(0,28 in)	33%.

(For larger thicknesses of members the Litzov curve allows an increase of the grain size.)

Mix proportions:

cement : aggregate : stabiliser : water $= 1:4,15:0,26:0,6$.

For an initial 28-day strength of 400 kp/cm² (5700 psi), this concrete still has a strength of 120 kp/cm² (1710 psi) after operating at 800 °C.

3.4.3.6. Compositions of a heat-resistant concrete made with copper slag for chimney elements of industrialised blocks of flats

The use of copper slag has now been proved on a commercial scale for the construction of the elements of chimneys, instead of the broken brick concrete which was formerly the normal material in industrialised flat construction [296].

The following constituents and mix proportions are required for 1 cubic metre of wet concrete:

230 kg   portland cement, class 350 (398 lb/yd³),
955 kg   broken raw copper slag 0 to 3 (2,5) mm (1655 lb/yd³),
955 kg   broken raw copper slag 3 to 7 (5) mm (1655 lb/yd³),
150,4 kg water (260,5 lb/yd³).

The mix proportions required to achieve a concrete quality of B 120 (1706 psi) are,

cement : aggregate : water $= 1:8,3:0,69$.

Better workability than that of broken brick concrete has proved to be one of the advantages of this material, in addition to the saving in broken brick and the good performance of this concrete in practice.

3.4.3.7. Composition of a heat-resistant concrete made with blast-furnace cement

The following composition has been used with success in the ceramic industries for a number of years for concreting the linings of the wagons in tunnel furnaces [97] (in weights per cent):

15% Blast-furnace cement, class 225,
10% Raw clay (from Haselbach) up to 0,75 mm (0,03 in) particle size (test point 1650 °C),
75% broken chamotte, 0 to 15 (12,5) mm (0,6 in).

Raw chamotte may also be used for the aggregate with equally good results. The grading, like that for concrete sands, should lie between the grading lines *A* and *B* of the standard TGL 0-1045. A mix of dry to stiff consistency (water/ cement ratio $\approx 0,7$) resulted in an initial 7-day strength of 120 kp/cm² (1710 psi), and a strength after burning at 1200 °C of 180 kp/cm² (2560 psi). On the basis of the values established for resistance to stress at high temperature ($t_a = 1140$ °C, $t_e = 1220$ °C) the temperature for continuous use of this concrete is given as approximately equal to 1150 °C.

3.4.3.8. Composition of a heat-resistant concrete made with smelted aluminous cement (trade name "Rolandshütte")

This mixture was used in the major repair of a brick ring furnace [137]. The operating temperatures came very close to the sintering limits of the burnt brick aggregate used; these limits lay between 1000 and 1100 °C. The grading of the aggregate followed closely that specified by grading curve *E* for normal (dense) concrete aggregates. The mix proportions were as follows:

    400 kg (881 lb) smelted aluminous cement, "Rolandshütte",
    500 litres (7,5 ft³) broken brick dust 0 to 3 mm (0,12 in),
    500 litres (7,5 ft³) broken brick 3 to 15 mm (0,6 in),
    200 litres (3,0 ft³) broken brick 15 to 30 mm (1,2 in).

3.4.3.9. Composition of a fire-resistant concrete made with SECAR cement

W. Grün [138] gives the following proportions for a fire-resistant concrete:

    95% silicon carbide, 0 to 5 mm (0,2 in),
    5% long asbestos fibres,
    410 kg of SECAR cement per cubic metre (710 lb/yd³) of compacted wet concrete,
    8% of added water.

In addition an incombustible plasticising agent which did not lead to air entrainment in the mix was used as an additive. The addition of the asbestos fibres results in a considerable improvement in the tensile strength in bending. With the figure of 5% quoted above an increase of about 50% can be achieved. The concrete described should, after subjection to a temperature of 1600 °C, have a compressive strength of more than 800 kp/cm² (11,370 psi).

3.4.3.10. Composition of a fire-resistant concrete made with smelted high alumina cement (Zschornewitz)

The smelted high alumina cement produced experimentally in the German Democratic Republic (72% Al_2O_3, 25% CaO) has been used in two mixes of fire-resistant concrete, which have both been tested in industry [297] (Quantities in weights per cent):

Corundum Concrete	*SiC Concrete*
20% smelted high alumina cement, Zschornewitz	20% smelted high alumina cement, Zschornewitz
30% corundum No. 8	30% corundum No. 8
25% corundum 200 or 250	25% SiC 200 or 250
25% corundum 100	25% SiC 100.

The water required for a dry to stiff consistency of concrete is given as 12 parts to 100% of weight. Based upon the tests of resistance to stress at high temperatures (for corundum concrete, $t_a = 1320$ °C, $t_e = 1730$ °C; for SiC concrete, $t_a = 1685$ °C, $t_e = 1710$ °C), the use of this concrete as a fire-resistant material is possible.

3.4.3.11. Composition of a heat-resistant concrete using additives

The example of a concrete for a chimney may be considered here. Most chimneys are not subjected to temperatures as high as those for the majority of structures which are built of concretes designed for high temperatures (their temperatures are usually in the region of 200 to 300 °C). Considerable variations of temperature are, however, of some importance in this case, and the concretes used must have the highest possible resistance to attack by sulphurous gases. The aggressiveness is considerably increased if the temperature drops below the dew-point, resulting in the formation of sulphurous acids or sulphuric acid. The use of porous aggregates, of which broken brick is an example, certainly gives good resistance to the flow of heat and ensures an equable distribution of temperature stresses, but the surface of the concrete is not as dense as it would be if normal concrete aggregates were used. The inner surface of the chimney is therefore sealed with waterglass, which will penetrate the pores and block them up. In addition borax is often applied afterwards, which assists further in the densification of the surface, absorbs particles of soot and sets hard. In the mix described by *W. Grün* which is quoted below [138], borax is included in the concrete mix. The choice of basalt and broken brick for the aggregates ensures a good combination of high density and low conductivity in the resultant heat-resistant concrete.

Aggregates:	Basalt dust	0	to	0,2 mm	(0,008 in)	8%,
	Basalt sand	0,2	to	1 mm	(0,04 in)	12%,
	Basalt sand	1	to	3 mm	(0,12 in)	15%,
	Broken brick	3	to	7 mm	(0,28 in)	15%,
	Broken brick	7	to	15 mm	(0,6 in)	18%,
	Broken brick	15	to	30 mm	(1,2 in)	15%,
	Broken brick	30	to	50 mm	(2 in)	17%;

Binding agent: Blast-furnace cement, class 275, in the proportion of 350 kg per cubic metre (606 lb/yd³) of wet concrete;

Water: 9%;

Additives: 2% plasticiser, incombustible compacting agents and $^1/_2$% of borax to the cement.

(Borax is also applied in the after-treatment of the surface.)

The cold compressive strength of the heat-resistant concrete made from this mix is given as 300 kp/cm² (4260 psi).

3.4.3.12. Composition of a fire-resistant concrete made with periclase cement

Using a so-called periclase cement and chrome ore-magnesia as a filler, a concrete has been developed which shows satisfactory stability in the presence of liquid, basic to slightly acid, blast-furnace slags [298]. The mix proportions by weight are:

periclase cement:finely ground aggregate:filler = 1:0,5:1,5.

A water/cement ratio of only 0,24 is required, because of the small water demand of the materials used. As stated in Section 2.2.2, the finest fraction of the magnesia aggregate is known to act as a binding agent when periclase cement is used.

The composition of a concrete of this type can be given as:

35 to 40% periclase cement,

60 to 65% chrome ore-magnesia filler (0,6 to 5 mm, and 5 to 20 mm).

Depending upon the particular design, the proportion of the fine to the coarse fraction will be 1:1 to 1:2. The resistance to temperature of this type of concrete is greater than 1710 °C.

4. Manufacture, Placement and Commissioning

Following the design and detailed calculation of the mix proportions for a refractory concrete, which are determined by the purpose which the concrete is intended to serve and in particular by the expected operating temperature, and which depend upon the type and quantity of the binding agent and aggregate, comes its manufacture. In principle this process follows the technology that is normal for general concretes for structural purposes. It must be noted, however, that there is a difference in that, under certain circumstances, difficulties can arise in relation to the workability of the mix, in the case of certain special aggregates which are not normally encountered in the majority of concrete work. One of the cases in which this can occur is when porous aggregates and those of very fine particle size are used for cement-bonded mixes. Similar difficulties can arise in the case of concretes bonded with waterglass, where the handling requires special attention.

The detailed discussions which follow, like those of the previous chapter, relate principally to the manufacture of high temperature concretes using normal industrial cements. These are the most commonly used and therefore provide the widest and most reliable field of experience. The examples studied here contain in part the principles of general concrete technology and in part some special aspects which are peculiar to refractory concretes. They relate to specific cases, but could give rise to different solutions when applied in other circumstances.

Information is becoming more available in Soviet literature about the use of reinforced concrete at temperatures in the middle range. The subject is, however, still relatively new, and detailed discussion of design and construction procedures should not therefore be undertaken until more quantitative experience is available.

The manufacture of lightweight concrete has many special features which differ from those of dense concrete. A number of these points will therefore be specifically dealt with when discussing normal concrete.

4.1.   Manufacture of Concrete

All the operations necessary to the preparation of wet concrete before it is given its final shape come under the heading of concrete manufacture. The most important of these operations comprise the proportioning and mixing of the constituents, that is the cement, aggregates and water, and the production of a mix which is capable of being worked into shape. However, since the properties which determine the workability are essentially a function of the water content, some of the problems raised by the taking up of water by porous aggregates and their place in the technology will be discussed before going on to consider the mixing process.

4.1.1.   Absorption of Water by Porous Aggregates

The ability of porous aggregate particles to absorb water must be known in advance and must be taken into account in deciding the proportions of the mix, if the correct consistency for proper workability is to be achieved. For example, some French instructions for the preparation of refractory concretes [121] require that the aggregates shall be sprayed with water and the surplus water then allowed to run off. The surface of the material should then be allowed to dry in air and the aggregate should be used on the following day. So long as this treatment is given to aggregates of a size larger than 5 mm (0,2 in), it may be efficacious in increasing the strength of the concrete. If it is applied to the finer fractions as well, however, it is either difficult to guarantee complete soaking of the material or else the fines entrain more water than is necessary for their saturation and thus give a false value for the water demand of the mixture. It must also be remembered that with this method of so-called "pre-spraying" the water/cement ratio can vary drastically and may even become virtually indeterminate.

The more common and more exact method is to moisten the aggregates directly in the course of preparing the mix. For porous aggregates it is then necessary to find out experimentally how long the largest particle takes to become completely soaked. This time must then be adopted as the period that is necessary for the mix to lie before it can be placed and compacted. This is the only way to ensure that the concrete will not be "thirsty". The standing time necessary for a porous broken brick aggregate, up to a particle size of 15 mm (0,6 in), may be quoted as an example; this takes 20 to 30 minutes, though this does not apply to concretes that are to be worked in a liquid state. However, smelted aluminous cement concretes, should not be allowed to stand for more than 20 minutes, otherwise the setting process may start; concretes made with portland cement should not stand for more than 1 hour. The periods are calculated from the start of the mixing process. One further advantage arises from the necessity to allow the concrete to stand for a period; this is that the later a concrete is compacted, the higher the strength which it finally reaches [138].

4.1.2.   Methods of Mixing

Hand-mixing, which may be carried out either on a mixing bench or on a levelled slab, is the oldest method, but should be considered for heat-resistant concretes only if a small quantity is required, as for example in the case of repair work. Mechanical mixing is normal in all other cases.

Rotating drum mixers and pan mixers are distinguished by their method of operation. Their size is described by the volume of dry material which they can contain. In the rotating drum or free-fall type the drum revolves, lifting up the materials to be mixed and dropping them from the highest point of revolution. In certain circumstances this type is suitable for heat-resistant concretes. The pan type of mixer has an advantage, however, if the mix is stiff and contains porous aggregates which are difficult to compact, despite the fact that this mixer requires a larger amount of power to drive it than the rotating drum type. The blades of the pan type mixer knead the materials so thoroughly that a shorter mixing time is required. A further advantage is that supervision of the mixing process is made easier by its being visible. The pan mixer is therefore particularly useful in stationary plants producing large quantities of concrete.

4.1.3. Mixing

The minimum mixing time specified in TGL 0-1045 [127] for normal concretes in rotating drum mixers is from 60 to 120 seconds. Mixing time in this context means the time during which the mix is being worked in the drum. The time should be longer for very fat mixes, say 90 to 120 seconds in a rotating drum mixer and 60 to 90 seconds in a pan mixer. It is stated that for normal concretes a mixing time greater than 180 seconds does not lead to any further increase in strength.

The achievement of thorough homogeneity in the mix of high temperature concretes, except for those made from non-porous aggregates, is rather more difficult than for normal concretes and therefore requires longer mixing periods. When portland cement or blast-furnace cement with a very fine micro-filler aggregate is used, the latter must be distributed as uniformly as possible to perform its function properly. Since such micro-fillers as raw clay or chamotte dust from rotating kilns soak up a large amount of water and then have a slight tendency to form lumps, there is an advantage to be gained from pre-mixing the dry constituents. Even by this means complete homogeneity of the constituents, particularly of the finest fraction, cannot be guaranteed. A still better method is to grind the cement clinker together with the micro-filler, to the required fineness.

The following mixing procedure, which has been developed in Poland [123] for high-temperature concretes having a portland cement base and has been used in rotating drum mixers, has proved very successful. The coarser fractions of the aggregate, larger than 2 mm (0,08 in), are weighed and fed into the mixer, together with one half of the water; these constituents are then mixed for 5 minutes at the full speed of the mixer. This ensures that the surfaces of the aggregate particles are fully wetted. The correct weights of cement, micro-filler and fine aggregates, together with the remainder of the water, are then added. The whole is then thoroughly mixed for 10 minutes. After a standing period of up to one hour, depending upon the water demand of the aggregate, concreting may be commenced. This procedure is particularly suitable for making small quantities of concrete on site.

In general, mixes for refractory concretes are delivered already proportioned in the dry state. It is then advisable to mix the constituents at the full speed of the mixer for a few minutes, while they are still in the dry state, in order to obviate the possibility of segregation, following up with the water, which should be added by stages.

4.2. Working of Concrete

The working of the wet concrete is to a great extent dependent upon its consistency. The compaction which is so important for structural work will be a function of this property; the relation between consistency and compaction therefore takes first place in the discussion in this Section. Amongst other things, some special aspects of reinforcement are also dealt with here. The working of concrete is also associated with questions of its subsequent curing, which unfortunately are often insufficiently considered, resulting in loss of quality in the final product.

4.2.1. Consistency and Compaction of Concrete

The factors that determine the degree of consistency or stiffness of a wet concrete are: the water content, the grading and shape of the aggregates, the cement content, and the quantity of micro-filler. *A. Hummel* [124] states that the consistency positively and directly influences only the workability and compactibility of the wet concrete and the type and uniformity of the final concrete's internal structure. Its effect upon the strength of the hardened concrete is only an indirect one, acting via the parameters just listed.

A distinction should be drawn between the following broad classifications:
(a) dry-to-stiff or stiff concrete, as tamped or vibrated concrete,
(b) plastic or soft concrete,
(c) liquid or pourable concrete.

The tests for measuring the consistency of concrete are the penetration test for stiff concrete and the slump test for soft or liquid concrete.* The standards required according to these tests are specified in the standard TGL 0-1048 [137a], for normal dense concrete. No special standards have yet been laid down in the German Democratic Republic for high-temperature concretes. Various methods have been specified in the Soviet Union for establishing consistency [8].

Placing by hand and handling by means of skip and hoist require various degrees of workability. The use of other mechanical placing methods requires particular ranges of workability.

4.2.1.1. Dry-to-stiff concrete

Dry-to-stiff concrete, if it is properly compacted, gives the highest strength. Hand compacting is normal for small quantities, but the maximum particle

---

*) *Translator's note:* These are tests current in the German Democratic Republic. The compacting factor and slump tests respectively correspond approximately to them.

size should not exceed 60 mm (2,4 in) in this case. The height of a lift must be established by acceptability tests.

A lift height of 20 cm (8 in) can be well compacted by machine ramming (either by the use of compressed air or by electricity), if the correct concrete stiffness is used, and a compacting time of 1,5 minutes per square yard or square metre is adopted.

Compaction by means of vibrators is faster and more reliable than compaction by ramming. For equal cement contents, vibrated concrete also achieves much higher strengths, as can be seen from Fig. 29 [14]. In addition there is no limit to the size of aggregate if vibration is used.

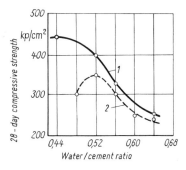

Fig. 29. Compressive strength of rammed and vibrated concretes for various water-cement ratios (Schulze)

1 vibrated; *2* rammed

The types of vibrator include internal (poker) vibrators, surface vibrators, those which operate externally on the shuttering and also vibrating tables. Those that operate at a frequency of between 3000 and 9000 per minute are the most satisfactory.

Internal vibrators of high energy are suitable for the larger masses of concrete. With this type it is possible to place stiff mixes to a high degree of compaction. According to *A. Hummel* [124], the stiffness of the mix is correct for a particular vibrator if the concrete closes up when the vibrator is withdrawn slowly. If hollows are left behind in the concrete by the vibrator, this indicates that the mix is too stiff and is not capable of being vibrated.

External or shutter vibrators may be applied either to the surface of the timber shutter or, in the case of precast units, by means of suitable attachments to the steel moulds. Since the effect of this type of vibrator does not penetrate very far into the concrete, it is limited to members of the slab type.

Surface vibrators consist of vibrating plates or planks. The concrete mix for this type should be slightly wetter than a dry-to-stiff mix and should be placed in a layer of uniform thickness. As an approximate guide, the depth placed should be equal to about four times the size of the maximum aggregate.

Vibrating tables are used chiefly for the manufacture of large numbers of precast units of small size. In the case of larger units it is very important to check the uniformity of compaction.

For making large numbers of smaller elements, presses are also particularly suitable, especially revolving table presses. Concrete with a very low water content can be compacted by this means.

4.2.1.2. Plastic concrete

Where a dry-to-stiff mix would be too difficult to work (especially in the case of complicated components or repair operations where access is difficult, such as in ovens), plastic or soft concrete is used. Mixes of this type have low strengths, but have an advantage over those which are difficult to compact, in that the surface is smoother and there is less risk of hollows forming. A more plastic mix does also exhibit greater age-hardening than a drier mix [138]. An assessment has to be made for every case, to decide whether the difficulties of compaction are balanced by the risk of bad construction. If a mix is too wet, vibrating will not improve the quality, since segregation is possible due to sinking of the aggregates and rising of the water. The water then takes cement with it, to the detriment of the internal structure of the concrete.

In principle, all the types of vibrator mentioned above may be considered for plastic mixes, but internal vibrators are to be preferred. Care must always be taken to ensure, however, that internal vibrators are not used to spread plastic concrete mixes, but only for compaction, otherwise severe segregation of the mix may result.

4.2.1.3. Liquid concrete

If it is important to have a concrete that is capable of being easily placed, liquid or pourable concrete will be used. This class of concrete is easily pumpable, but has a strong tendency to segregate if not properly handled. There are therefore limits to the particle size of aggregates and fine fractions. A concrete of this type should not contain too high a proportion of the coarse fraction, in order to ensure proper cohesion; on the other hand the fines content should not be too great, or the water content will need to be raised still further. As far as possible, the use of normal cements and micro-fillers should therefore be avoided in liquid concretes. In special cases the use of pourable concrete, preferably made with aluminous or high alumina cement, may be authorised, since these cements have a higher heat of hydration because of their shorter setting time. The repair of plant which is still warm is a special case for the use of liquid concretes, since the excess water has the effect of ensuring that the concrete surface does not dry out and break up. The use of liquid concrete may also be unavoidable in places to which access is difficult, because stiffer mixes may be incapable of proper compaction under such conditions.

A special type of concrete to be noted is sprayed concrete, which is shot by means of compressed air on to the shuttering or on to concrete which is already hard. The Torkret system is used to a considerable extent in repair work. *H. Zahlbruckner* [139] has described the excellent results obtained when concreting oil-burners with perchromite refractory concrete by this method. A dry mix of cement and fine aggregates, up to 3 mm (0,12 in), is used. The proportion of water added is determined by the mixing apparatus of the Torkret gun and should be kept as low as possible consistent with the adhesion of the concrete. The jet must be maintained as nearly as possible in the perpendicular position to avoid excessive loss due to rebound. The material which does fall away should not be used again, as it is a segregated product. Better adhesion can be ensured over large areas and where the thickness is greater than 30 mm (1,2 in) by the addition of mesh reinforcement, provided that the subsequent operating temperatures permit it. Concrete applied in this way is very well compacted.

The disadvantage that the water content is not always well defined is overcome by the use of the so-called Moser sprayed concrete system, in which the mortar is combined with water right at the start and is forced in a moist condition through a short compressed air tube. With the Torkret system on the other hand the length of the tube can be varied within limits.

4.2.2. Reinforcement

Reinforcement in refractory concrete has two principal functions: (a) to resist static stresses, and (b) to prevent cracking. In normal reinforced concrete the reinforcement and the concrete itself behave in a compatible manner; that is in normal circumstances they both undergo the same expansion. They behave quite differently, however, at high temperatures. At a temperature of about 400 °C steel reinforcement suffers an appreciable loss of strength. In addition the effect of the differences in the two coefficients of thermal expansion becomes much more marked at high temperatures.

In the temperature range between 100 and 250 °C the bond strength between steel and concrete increases uniformly with the concrete strength. At higher temperatures the expansion of the steel increases, while the strength of the concrete decreases. Plain round bars, for instance, lose 75% of their bond at 450 °C, while ribbed bars have the same bond strength at 450 °C as at 20 °C [122]. It is therefore advisable to use ribbed bars of cross-section greater than 4 mm (0,16 in); steel of smaller diameters is more easily damaged by oxidisation, which can be attributed to the formation of cracks in the concrete and which leads to a volumetric increase.

The use of cast iron and particularly of special-purpose heat-resistant steels is also well known [139]. Even so, it is nevertheless important to arrange the steel in the outer, lower-temperature, layers of the structure. The use of I-beams, partly embedded in the concrete and partly projecting from it, is also possible. This results in an improved dissipation of heat. This type of structural arrangement should be avoided, however, if the operating temperature is greater than 1000 °C.

The need to allow the steel to expand freely at high temperatures has led to the evolution in different countries of various methods of ensuring that there is a small space between steel and concrete. These include the coating of the bars with a thick layer of grease, or surrounding them with a sheet of paper, tar or bituminous material of a thickness equal to at least $1/3$ of the diameter of the bar. After firing, the reinforcement has a clear space available in the concrete in which it can move, and there is no danger of its bursting out.

Fig. 30 shows a design solution for a flat element with I-beam reinforcement [6]. It can be seen that angles are fixed to the I-beams. The angles are cut into sections and bent over alternately on opposite sides, this improving the bond to the concrete (a). The upstanding portion of the beam serves as a support member for the side of the furnace. In this design the beams are embedded in a layer of insulation after erection (b).

Figs. 31 and 32 show examples of roof and wall construction for boilers. The roof component is of hanging construction. Here the rolled steel section rests on the wall of the boiler and steel stirrups welded to it support the concrete component below. The reinforcement of the latter is from 6 to 8 mm (0,24 to 0,32 in) in diameter. The space between can be filled with mineral wool or

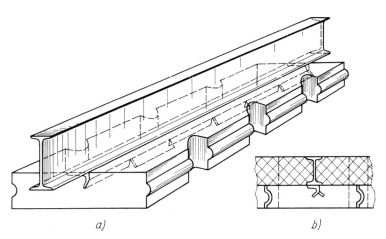

a)   *b)*

Fig. 30. Pre-cast units heat-resistant concrete reinforced with steel I-beams (Nekrasov)

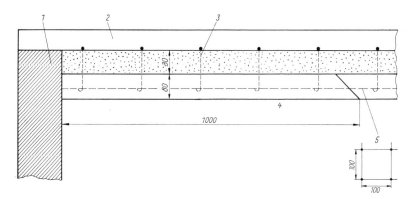

Fig. 31. Suspended roof with precast units of heat-resistant concrete (Dimensions in mm)

1 Wall of boiler; *2* Rolled steel section; *3* Steel hanger or stirrup, 6 mm ($^1/_4$ in) diam;
4 Heat-resistant concrete; *5* Reinforcing steel, 6 mm ($^1/_4$ in) diam

other suitable insulating material. The anchored and partly concreted-in rolled steel sections of these precast wall components enable the components to be fixed firmly to the steel wall framing of the boiler.

In Fig. 33 can be seen the design that was evolved for the drop gate of a large chimney flue, designed in accordance with the standard TGL 9233 [140]. This was manufactured from heat-resistant concrete in the VEB Silica and

Chamotte Works at Rietschen. The suspension lug by which the gate is lifted is attached to the steel reinforcement. This gate has proved very satisfactory in service. If the gate section is damaged, the lifting beam can be bolted on to a new section.

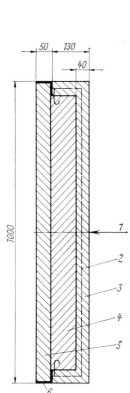

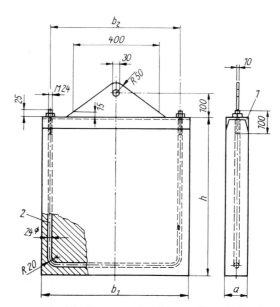

Fig. 33. Drop-gate for chimney flue, designed to TGL 9233 (Dimensions in mm)

1 Steel channel section to TGL 7960; *2* Steel U-holt

Fig. 32. Reinforced precast unit of heat-resistant concrete for boiler walls (Dimensions in mm)

1 Hot side; *2* Reinforcing steel, 6 mm ($^1/_4$ in) diam, as Fig. 31; *3* Heat-resistant concrete; *4* Insulating material; *5* Heat-resistant concrete; *6* Rolled steel angle 50/50/5 (approx. $2 \times 2 \times ^1/_5$ in)

4.2.3. Shuttering and Forms

In general the principles which apply to shuttering and formwork for refractory concrete are the same as for ordinary concrete. Only certain special aspects of this subject for refractory concretes will therefore be discussed here. General rules for the manufacture and erection of shuttering are given in the standard TGL 0-1045 [127].

Timber shutters are still the most commonly used, because of the simplicity of their manufacture and their relatively low cost. If rough timber shuttering is used the visible face of the concrete is usually not very smooth. Timber has a tendency to soak up water and therefore to draw it out of the concrete. Small particles of cement also become detached and stick to the rough surface of the shuttering. The formwork is therefore coated with mould oil to prevent this and should also be wetted immediately before concrete is placed.

Undesirable effects at the interface of concrete and formwork can be avoided by laying sheets of synthetic material such as plastic between the concrete and the form or by covering the latter completely with the material [14]. In addition to making the formwork more durable as a whole, this procedure enables economies to be made in the use of timber. An alternative is to face the timber shuttering with steel sheet. In this case the surface still needs to be oiled.

The use of steel shuttering and forms is continually on the increase. Since the properties of steel are less variable and its modulus of elasticity is much higher than that of timber, relatively thin steel plates may be used for shuttering [14]. Steel plating is usually stiffened by a backing of rolled steel sections or, in the case of large areas of formwork, by full steel framing. Oiling of the face of the shutter is still absolutely essential, to prevent the concrete from adhering to the steel and to prevent the latter from rusting. The disadvantages of steel forms and shuttering include the high cost of initial manufacture and the greater weight. The advantages can outweigh these disadvantages, however, especially where precast units are to be manufactured and where a high standard of accuracy and of surface finish is required. Steel shuttering has the advantage over timber, that it is much better in locations where mechanical compaction of the concrete is to be used.

When high quality aluminous cement is employed, the concrete must be sprayed with water from the commencement of hardening, that is to say from 4 to 5 hours after pouring. The shuttering has therefore to be designed so that not more than, 2 hours is necessary for concreting, inclusive of the time for mixing.

4.2.4. The Construction of Joints

It is evident that it is often necessary, due to the variable strain and shrinkage characteristics of concretes, to provide joints in a structure for high temperature purpose. Their extent will depend upon the dimensions of the structure and the type of concrete to be used. This question arises particularly when large structures have to be subdivided for the sake of convenience, such as in the case of the wagons for a tunnel furnace, or when a wall is to be erected from a number of precast units.

If suitable mix proportions are used, however, the need for joints can usually be avoided, since the thermal expansion of the concrete is cancelled by the shrinkage when it is first heated up. If the expansion and shrinkage characteristics of the concrete intended for a particular design are not known with any degree of accuracy, then suitable tests must be undertaken to determine them. It can then be determined in advance whether joints are necessary and how wide they should be.

In the case of precast units the arrangement of the joints is generally predetermined by the need to make up the joints with fire-resistant mortar. This mortar will have the ability to resist most of the stresses that arise. In high temperature furnace linings that are cast in situ care must be taken, when laying out the joints of the roof and abutment walls, to see that the corners of the joints are not subjected to opposing forces due to the working of the concrete under the effects of heating and cooling; otherwise cracking will occur.

The cast-in-situ section of the combustion chamber of an annular furnace is shown in Fig. 34 as an example of the layout of joints. This layout is straight-forward. Those which run horizontally in the outer structure of the roof and in the roof itself are designed to be only a few millimetres wide. This is quite sufficient if wet concrete is laid next to matured concrete. The concrete must be cast in such a way that there is no bond between the adjacent panels of concrete. In the transverse joints there is a distinction between those that run into the doors and those that pass either side of the doors. The former are not very wide (in this case 4 to 5 mm (0,16 to 0,2 in)) while the latter are 10 mm (0,4 in) wide.

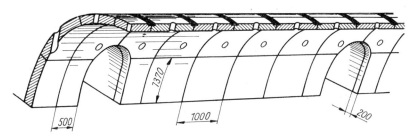

Fig. 34. Arrangement of horizontal and vertical joints in the cast-in-situ lining of the combustion chamber of an annular furnace (from [6]); Dimensions in mm

Timber or plywood may be used for forming joints of a particular width. Cardboard, corrugated cardboard, roofing felt and paper are also often employed. In large structures these materials can, however, have a detrimental effect on the strength, since they often cannot be laid straight and to a level. In addition they are often displaced by the compacting of the concrete, resulting in an irregular surface on the finished concrete. All "jointing materials" of this type are burnt away upon first firing, leaving a space which can be filled with fire-resistant mortar and pointed. It is particularly important to ensure that the concrete immediately next to the joints is very well compacted, to avoid corrosion being initiated in these particularly vulnerable positions.

The type of mortar most commonly used in joints is a normal chamotte mortar, consisting of 20% by weight of cement and 80% of chamotte graded 0 to 1 mm (0,04 in). For good workability this maximum particle size should not be exceeded. It is advisable to use as aggregate in mortars and pointing the same material as is used in the adjacent parts of the structure. The mortar is not too stiff for workability; if porous aggregates are used they should be wetted beforehand. The same is true for the joints themselves.

7*

4.3.   Hardening and Curing

The chemical hardening process commences at a certain interval of time after the mixing and compaction. The subsequent treatment is therefore of supreme importance to the achievement of the desired strength. The quality of the finished concrete product is very much influenced by this treatment, as well as by the proportions of the mix itself. While setting is taking place care must be taken to ensure that the hardening concrete is not subjected to any vibration, otherwise the concrete structure may develop cracks. These cracks may subsequently close up but nevertheless they will have an adverse effect upon the concrete. Mechanical vibration can lead to such serious cracking, in an advanced stage of hardening, that the cracks will never disappear and the resultant damage to the structure of the concrete is extensive.

4.3.1.   Achievement of Early Strength

In the climate of Central Europe the range of temperature in which concreting is normally carried out is between $+5$ and $+30\,^{\circ}\mathrm{C}$. The temperatures specified for laboratory research and test work are $+18$ to $+20\,^{\circ}\mathrm{C}$. The processes known as normal hardening take place in these ranges of temperature.

In order to remove the shuttering earlier, efforts are often made to shorten the time necessary for hardening and thus to achieve a much higher early strength. Artificial measures of this type are termed accelerating measures. There are a number of possible procedures.

If an acceleration of the setting process is to be brought about by chemical means, then setting accelerators, as described in Section 2.4.3, may be used, after suitable preliminary testing to establish the correct quantities to be added. Another relatively simple method is to heat the concrete, without raising the moisture content of the air. In this case steps must be taken to prevent loss of moisture. Another possibility is steam curing, which has been used very successfully on high temperature concretes [17].

The reactions of different types of cement at high temperatures vary considerably. Iron portland cement and blast-furnace cement are very much affected. Steam hardening is not possible for smelted aluminous cement and super-sulphated cement [138], since at temperatures above $25\,^{\circ}\mathrm{C}$ these cements exhibit considerable loss of strength. Aluminous cements should hardly ever be worked at temperatures in excess of $+23\,^{\circ}\mathrm{C}$, on account of their marked sensitivity to temperature.

There are two types of steam curing: low pressure steam curing (between 30 and $90\,^{\circ}\mathrm{C}$) and high pressure steam curing (between 100 and $180\,^{\circ}\mathrm{C}$). The latter process is carried out under pressure in autoclaves. If this method is used, better strengths are obtained with quartzitic aggregates than with other types, since the reaction known as the lime-sandstone reaction results in the formation of additional tobermorite-like phases from the SiO_2-rich material and the free calcium hydrate of the hydrating cement. This phase-building leads to the development of strength. Reactions of this type do not start until a temperature in excess of $100\,^{\circ}\mathrm{C}$ is reached.

Low pressure steam heating for the accelerated hardening of high temperature concretes offers many possibilities for the future. It is sure to prove extremely economic in the manufacture of precast units. In addition to achieving quicker deliveries, the sizes of the areas necessary for storage can be kept smaller. For normal concretes a curing time of about 5 hours at a temperature of 70 °C should be satisfactory; 70% of the 28-day strength is usually reached after 12 hours of steam hardening at about 80 °C. The optimum figures for high temperature concretes have not yet been ascertained. As well as accelerating the hardening process the steam treatment reduces the tendency of the concrete to shrink. When the concrete units are removed from the steam curing room they should on no account be subjected directly to the normal atmosphere, since the rise of temperature that takes place in the immediate neighbourhood of the concrete causes the relative humidity of the air to have a very low value. This leads to rapid evaporation of the water, resulting in accelerated drying of the concrete and thus to shrinkage cracks. Furthermore, the hardening process should continue, following upon the steam treatment, and therefore the concrete units ought to be sprayed with water until they are fully cooled. After they have cooled, further wetting is unnecessary.

4.3.2. Wet Curing

It is necessary to cure all concretes, other than those that are subjected to steam hardening, by wetting them, in order to prevent shrinkage cracks developing and to ensure that the hydration reaction in the cement is not stopped by premature drying out. This is a fundamental requirement for all concretes, including special concretes for high-temperature purposes.

The wet curing of high-temperature concretes having a portland cement or blast-furnace cement base should be commenced 24 hours after concreting and should be continued for at least 7 days. In aluminous cement concretes the hydration process begins earlier and is more concentrated. Their curing should therefore start not more than 4 to 5 hours after placing. If high alumina cements are used the shuttering may be removed at this stage, except for highly loaded structures, thus making it possible to wet all the surfaces [121]. This is especially important in this case, since the concrete reaches a higher temperature than with normal cements, due to the shorter hardening time, and therefore more moisture evaporates. It is advisable to continue the curing of smelted aluminous cement concretes for three days.

It is preferable, wherever possible, to cover the concrete elements or structures that are to be cured with wet sawdust, sand or matting, rather than to spray them directly with water. Straw matting is suitable if the ambient temperature is low, as it has a good insulating effect. It is also possible and indeed advisable to utilise a covering of thick polyethylene sheets in strong sunshine or high winds.

Substances known as curing agents have been introduced in civil engineering practice to inhibit fast drying out of the concrete [14]. They may also be used with advantage in some cases in the making of heat-resisting concretes. The water content of the wet concrete is quite sufficient to ensure full hardening, provided that loss of water is prevented. Agents of this type are paraffin based or resinous substances. They may be used with portland or blast-furnace cement concretes and are applied by scattering in the form of a fine dust over the surface of the concrete. This should be done 20 to 40 minutes after the last vibrating of the concrete, or else after the surface becomes smooth. Surfaces which cannot be reached at this stage should be treated as soon as the shutters are removed.

In this connection reference should be made here to the excellent water holding properties of clays, in relation to the moisture in the surrounding air. In certain clays the water content can reach a value five times greater than the mass of the clay itself. If, therefore, these clays are used as the micro--filling agent they also have the effect of slowing down the drying process.

4.4.   Drying and Commissioning

The mechanical and physical properties of concretes that are going to be subjected to high temperatures are in great measure determined by the drying and firing procedures. If the correct rules are not observed right up to the first time that the operating temperature is reached, the quality of the concrete may be seriously impaired.

Though the method of drying and first heating will vary from one structure to another, it is a fundamental rule that these operations must not be carried out until the required strength of the concrete has been reached. The compressive tests are carried out on normally hardened test cubes after not less than 3 days in the case of aluminous cement concrete and not less than 7 days in the case of concrete made from portland or blast-furnace cement.

For the fire-resistant concrete mixes supplied by the VEB Chamotte Works at Brandis, which have a basis of smelted aluminous cement or ordinary portland cement, the following rates of temperature increase are laid down as applying to structures with a thickness up to 25 cm (10 in):

$$< 200\,°C \quad \ldots\ldots\ldots\ldots\ldots\ldots \quad 20 \text{ deg. per hour,}$$
$$200 \text{ to } 400\,°C \quad \ldots\ldots\ldots\ldots\ldots\ldots \quad 10 \text{ deg. per hour,}$$
$$400 \text{ to } 600\,°C \quad \ldots\ldots\ldots\ldots\ldots\ldots \quad 20 \text{ deg. per hour,}$$
$$> 600\,°C \quad \ldots\ldots\ldots\ldots\ldots\ldots \quad 100 \text{ deg. per hour.}$$

The rules drawn up in the U.S.S.R. for furnace aggregates lie in similar ranges, as indicated in Table 9 [6]. Here the term "small furnace" implies a furnace such as a reheating furnace, with a floor area up to 6 sq metres (about 65 ft²), and "large furnaces", such structures as tunnel or annular furnaces.

Table 9. Soviet regulations for the drying-out and raising to temperature of furnace plant (Nekrasov)

Heating stage	Type of furnace		
	small	medium	large
Raise temperature to 150 °C [deg/hr]	20···40	10···20	5···10
Holding time at 150 °C [days]	0,33	3,0	7,0
Raise temperature from 150 °C to operating temperature [deg/hr]	100···200	50···100	25···50

In the particular case of portland cement concretes, a Polish paper quotes the heating of a furnace up to a temperature of 110 °C in 10 hours, i.e. at a rate of about 10 deg. per hour [123]. This temperature was held for 24 hours, and was then raised to 600 °C at a rate of 50 deg. per hour. From 600 °C onwards the temperature can be increased at about 100 deg. per hour. The average time needed for bringing to an operating temperature of 1150 °C, which may be reckoned as the limit for portland cement concrete, is therefore about 50 hours.

Aluminous cements are especially suitable for use as hydraulic binding agents in fire-resistant concretes that have to harden quickly. It follows that these concretes will also withstand a more drastic raising to temperature. The information given in references [63] and [121] indicate that heating may commence as early as 24 hours after the concrete is poured, reaching a temperature of 500 °C at a rate of 50 deg. per hour. Heating beyond this point may be as fast as desired.

4.5.   Concrete Repairs and Surface Protection

Concrete structures and working parts that need repair may be either of concrete that has already been in service or of brick or other refractory lining materials. In many cases it may be possible to install prefabricated components; when this is not so, concrete must be placed in situ.

The use of precast units is therefore preferable when repairs have to be carried out to an operating furnace. This is often the case in brick-lined furnaces, in which the lining would be incapable of withstanding the temperature variations resulting from a sudden wetting. It must also be remembered that hardening of the newly-placed concrete would take place under abnormal and unfavourable conditions, resulting from the temperature of the structure. If, however, it is impossible to avoid the use of cast-in-situ concrete, smelted aluminous cement should be employed in almost all cases in order to keep the repair time as short as possible. This applies, whether or not the damaged section was made of this material. An almost liquid concrete should be used, especially if the brickwork is still warm, on account of the increased evaporation that will take place and to allow for the excessive absorption of water by the adjoining porous surfaces. The Torkret spraying system is very suitable for this type of repair, as the force of impact of the concrete as it is shot from the gun ensures a better bond to the existing structure. The bond can also be improved by using a mixture of equal parts of smelted aluminous cement and water for the pre-wetting of the structure [141].

Before any concrete is applied to them the brickwork or other structures to be repaired must be thoroughly cleaned and either roughened or chiselled, so as to offer a better bonding surface to the new concrete. They should then be wetted.

For repairs which for operational reasons have to be undertaken in the presence of considerable heat, a high-temperature-resistant mixture marketed under the name of "Thermobeton" is available [141a]. This mixture is compounded with water, thoroughly kneaded and will harden shortly after exposure to the air. It can be used for repairs to dry refractory brickwork or concrete components, without the need to remove slag and glazing from the surface which is to be repaired.

Certain materials known as "plastics" are also suitable for use in hot repair and improvement works. These substances have been developed cooperatively by the VEB Silica and Chamotte Works of Rietschen and the VEB Alloy Steel Works of Freital. This ceramic material exhibits only small shrinkage when it is heated, because the shrinkage of the clay is offset by the growth of the material used as the filler.

In this connection, those coatings and renderings should also be included here which are designed to give a certain amount of protection to the surface of refractory concrete. A substance known as "Schutzglasurmasse SM" (literally: basic protective glaze) has been developed [142], which has good bond and elastic properties and will provide a refractory structure with a long-lasting densified surface. There is an advantage in the reflective surface of coatings of this class, in that a large proportion of the radiant heat is thrown back into the combustion chamber and the heat retained in the furnace lining is consequently reduced.

From the same source there is also a material known as "Schlackenabweisende Schutzanstrichmasse SMC" (literally: slag-repellent protective coating). This is extremely resistant to attack by slag and forms a repellent surface coating which prevents slag adhering to it.

Coatings which are resistant to slag attack have also been developed in the Peoples' Republic of Poland. They are composed of waterglass, sulphite liquor, and broken glass from window panes or carborundum. The particle size lies in the range 0 to 1 mm (0,04 in), 70% by weight being in the 0 to 0,05 mm (0,002 in) fraction. Their expansion up to 1200 °C is low and, if carborundum is used, the shrinkage from 1200 °C is slight.

4.6.   The Manufacture of Cement-bonded Lightweight Refractory Concretes

It is opportune, after discussing the manufacture of dense concretes for high temperature purposes, to consider the making of lightweight refractory concretes. The technology of the latter differs to a certain extent from that of the former, so that a separate section on lightweight concretes is justified. In this case also, only the cement-bonded concrete can be dealt with, since most of the practical experience lies in this field.

Lightweight refractory concretes may be divided into two principal groups:

(a) Refractory concretes made with lightweight aggregates, and
(b) Porous concretes.

The two types differ appreciably in their manufacture.

4.6.1.   Refractory Concretes Made with Lightweight Aggregates

These comprise lightweight concretes having densities in the range 1,0 to 1,6, made with porous aggregates.

4.6.1.1. Refractory lightweight aggregates

In Section 2.3.1 the various types of lightweight aggregates and their applications in the manufacture of heat-resistant concretes are discussed. The following relates only to certain special aspects of their granulometry and some other specific questions.

For a porous and fairly firm structure, the best grading lies in the range 7 to 15 mm (0,28 to 0,6 in). The shape factor and surface properties of the particles will depend upon the origin of the aggregates and these have a decisive influence upon the working of the wet concrete and the amount of cement required. A compact shape has advantages, while flat or long particles are very unsatisfactory. The surface may be either of open pores or closed; either one or the other may be preferable, depending upon the use to which the final concrete is to be put. Aggregates which have open surfaces have proved better where high strengths are required. Those synthetic or artificial aggregates that have substantially acquired their final form before sintering, such as ceramic aggregates, have a hard structure and low water absorption. Aggregates such as blast-furnace pumice, sintered pumice, and boiler slag, which are crushed after sintering, have open-pored, fissured structures and varied shapes; consequently the method of working them is different. Thus the aggregates of the first group give good workability and high strength with a relatively small proportion of cement and water and also do not shrink excessively, while those of the second group require much higher proprotions of cement if equivalent strength is to be achieved. In addition, concretes made from the second group of aggregates usually have greater shrinkage and are more difficult to work.

The strength of the separate particles will depend principally upon the structure and nature of the material and upon the pores contained within it. The mechanical strength of a particle is considerably reduced by the presence of very large pores, which typically may reach as much as 20 mm (0,8 in) in diameter. Particle densities vary considerably with different materials; there are also large variations within the same fraction due to differences in the porosities. Thus expanded substances, where the fine fraction consists essentially of the same material as the coarser particles, possess higher particle densities and lower porosities as the particle size becomes smaller. Miscellaneous crushed aggregates composed of material of many different types, on the other hand, have higher porosities and lower particle densities at the finer end of their grading range. These fines arise from the crushing of the larger, easily-crushed pieces, while the well-sintered particles offer the greatest resistance [15]. Ceramic aggregates produced by a burning process have a fairly uniform structure and hence almost invariant particle density throughout the grading range. It must be remembered, however, that if a large proportion of fine aggregate is used in a concrete, a relatively high overall density will result.

The same rules apply in general, with regard to water absorption or take-up, as have been stated for normal heat-resistant concretes made from porous aggregates. The amount of absorption depends not only upon the pore volume, but also upon the size of the separate pores. The highest absorption occurs when there are very fine pores and capillaries.

The ordinary rules apply with regard to the permissible proportion of defective aggregate particles. Particular attention should be paid to the SO_3 content, which must in no case be allowed to exceed 1,5% by mass. It is especially important to carry out a test analysis for this purpose in the case of aggregates made from boiler slag.

4.6.1.2. Composition of refractory concretes made with lightweight aggregates

The porosity of a lightweight concrete is a function of the "self-porosity" of the aggregates (that is, the porosity of the aggregate as a material), the shape-factor and surface quality of the aggregates, and also of the porosity of the whole concrete mass. The aim is to utilise these porosities to the best advantage in obtaining good insulation properties, while also achieving the required strength. The grading rules for normal concrete are, therefore, not generally applicable to aggregates for lightweight concretes. Consequently, grading curves may be used only to a limited extent. The most important factors in selecting the separate fractions are the size of the aggregates and the nature of their particles.

Concretes made with lightweight aggregates are usually divided into the following three groups:
(a) Lightweight concrete having porosity in the particles of the aggregate. This implies closed porosity, with hollows inside the particles.
(b) Lightweight concretes with porosity in the mass. The open porosity is provided by the spaces between the aggregate particles.
(c) Lightweight concretes possessing both particle porosity and mass porosity.

The lightweight concretes of group (a) possess good insulation properties and also have structural strengths that are adequate for load-bearing; for groups (b) and (c), on the other hand, insulation is the most important property. Usually when porous aggregates are used, a form is chosen which gives mass porosity as well, the fines of the 0 to 1 mm (0,04 in) or the 0 to 3 mm (0,12 in) range being sieved out. This procedure gives rise to a large number of air pockets between the particles, while at the same time there is a reduction in the quantity of cement required. Gap grading is another possibility. The following mix proportions have been found to give a satisfactory level of mass porosity together with adequate strength [15][1]:

> 30 to 40% of aggregate 3 to 7 mm (0,12 to 0,28 in),
> 70 to 60% of aggregate 10 to 15 mm (0,4 to 0,6 in).

If a higher proportion of fines is used, the increase in compressive strength is accompanied by an undesirable increase in density.

Very fine additives, such as kieselguhr dust or brick dust, may also be used to help in achieving a normal setting reaction while maintaining the relatively low cement content of lightweight concretes. Researches in this connection by *A. Gburek* and *R. Röhlig* [310] with pearlite concrete have demonstrated that if very fine additives are not present there is a marked delay in the setting process.

Since the aggregate strength is a decisive factor in determining the strength of a lightweight concrete, a high cement strength has an effect only if the

---

[1]) The gradings are given here in percentages by volume, since we are concerned with the volume displaced.

aggregate is a strong one. *A. Hummel* [124] gives the cement contents of a wide range of lightweight concretes as from 100 to 300 kg of cement per cubic metre (168 to 504 lb/yd³) of wet concrete. The upper limit in the case of heat-resistant lightweight concretes is also 300 kg per cubic metre. Thus in the case of pearlite concrete, when the cement content was increased above 300, the compressive strength continued to increase only slightly up to a limit of about 40 kp/cm² (570 psi) and the conductivity deteriorated so much that the material was no longer economic as an insulating concrete [301].

Because of the numerous variations in the properties of the aggregates, the correct quantity of water can be determined only by making trial mixes and testing their consistency. In addition to allowing sufficient water for absorption by the porous aggregates, care must be taken to ensure that the cement is combined with enough water to cover completely all the particles and penetrate into crevices. Where practicable, plasticising agents should be used, to avoid the need for excessive water contents. This is especially true where pearlite, which is extremely water-absorbent, is used as the filler material. Concretes of this type may require a water/cement ratio of up to 1,1.

Reference should be made to Section 4.1.1 in regard to the water required, when the proportions of the mix are being determined. *S. Reinsdorf* [15] recommends that in the manufacture of concretes from lightweight aggregates, the cement and water should be proportioned by weight and then related to the volume of the aggregate, loosely compacted, or to the volume of the final, set, concrete. Thus the mix may be stated:

Mix proportion (wet) = MP_{wet} = kg (or lb) of cement:m³ (or yd³) of air-dried, loosely packed, aggregate:kg (or lb) of water

and for the purpose of comparing quantities of cast concrete:

Mix proportion (cast) = MP_{dry} = kg (or lb) of cement:m³ (or yd³) of cast concrete (air-dry).

4.6.1.3. Working and subsequent treatment

Reference should be made to the sections on normal high temperature concretes and to technical literature on lightweight concretes for these aspects of the subject. Only a few principles will be stated here.

Pan mixers are the type usually employed, since, quite apart from the advantages of this type of plant, there is a tendency in the rotating drum type of mixer for porous aggregates to be broken up by the falling action and to be scoured away by the rubbing of the drum.

In general the result of intensive compaction of a lightweight aggregate concrete is to nullify the effort to obtain a light density. On the other hand, a concrete made from a lightweight aggregate that is difficult to work must be compacted. In principle it should be lightly punned; structural elements will require rather more compaction.

Concretes made from lightweight aggregates can be induced to develop their strength at an early age by the usual method of hot curing. Hot water curing, however, is not a correct form of treatment. While wet steam and hot, dry air are acceptable curing media for concretes with aggregates that are only lightly absorbent, in the case of aggregates that are highly absorbent wet steam treatment is harmful, as the researches of *Reinsdorf* into broken brick concrete have shown [15]. The lower the moisture in the heat treatment chamber, the better the strength, both initial and final. The particles of the aggregates contain enough moisture to ensure that the strength is developed, when hot dry air is used.

The good effects obtained from heat treatment result from the fact that the initial strength increases with increasing porosity of the concrete. The subsequent gain of strength is admittedly somewhat reduced. Also, the more porous the structure of the concrete and the higher the temperature of curing, the less will be the subsequent development of strength. Expanded shale concrete may have an increase of strength, up to 28 days, of 20 to 40% and broken brick concrete, 25 to 30%.

With regard to subsequent treatment, current specifications state that lightweight concrete should be kept moist for at least 3 days; in the case of impermeable shuttering the period may be reckoned from the time of removing the shutters. *O. Graf* [143] has shown from tests on broken brick concrete that less care need be exercised during the curing, on account of the considerable quantity of water that is absorbed and stored. Cubes hardened in the open under unfavourable conditions exhibited the same strength as those which were stored either with or without any subsequent treatment, in the laboratory.

4.6.1.4. Lightweight heat-resistant slag wool concrete

Slag or mineral wool as an aggregate possesses certain special features when compared with the lightweight aggregate concretes so far considered. This type of concrete is therefore now considered separately.

The fibres of mineral wool contribute to the structure of a wet concrete by forming a kind of stable framework; when the cement sets they also contribute somewhat towards its strength. They reinforce the component parts of the material and prevent them from being damaged by the stresses caused by volumetric changes. They also improve the bending, tensile, and impact strengths of precast units. Practical experience has been obtained in the use of this concrete in the insulation of plants up to temperatures of 900 °C [299].

A. F. Milovanov and *V. M. Prjadko* [100] have reported the use of slag wool obtained from blast-furnace slag and having a density of 250 to 350 kg/m³ (155 to 21,8 lb/ft³). The binding agent was portland cement of the Soviet quality 400, or waterglass of a density of 1,38 g per millilitre (72 lb/ft³) and a module of 2,5. Chamotte powder was used as the ceramic stabiliser to improve the heat-resistant qualities of the concrete. In the waterglass concrete, sodium silico-fluoride was also added, in the proportion of 9 to 10% of the waterglass content.

Using portland cement, the following mix proportions were tested (quantities in weight per cent):

cement:chamotte dust:mineral wool = 31:15:54.

The quantity of cement required was 200 kg/cubic metre (336 lb/yd³) and the water/cement ratio was about 2. In mixing, the slag wool was first combined with from 30 to 50% of the water in a concrete mixer and mixed into a paste. Then the cement, previously mixed with the chamotte dust, was added to the mixture. The rest of the water was added at the same time. The whole was thoroughly mixed for from 5 to 7 minutes and was then ready for use.

With waterglass as the binding agent the following proportions are recommended (quantities in weight per cent):

waterglass:Na_2SiF_6:chamotte dust:mineral wool = 37:4:12:47.

The quantity of waterglass required was 300 kg/cubic metre (504 lb/yd^3) of concrete. In this case also a mixer was used. The slag wool was first combined with 30 to 50% of the waterglass. The chamotte dust and the Na_2SiF_6 were then added, followed by the rest of the waterglass, the whole being subjected to thorough mixing. Mineral wool concretes of this type have densities after drying out at 110 °C of 700 to 800 kg/m^3 (43,5 to 49,7 lb/ft^3) when made with portland cement, and 800 to 900 kg/m^3 (49,7 to 55,9 lb/ft^3) when made with waterglass.

4.6.2. Porous Concretes

Reinsdorf quotes the following description of porous concrete [144]: "An artificial stone-like material, having a very porous, uniform structure, in which the pores are connected with one another by fine openings and capillaries. Porous concretes are made from fine-grained mineral materials, hydraulic binding agents, water and a special additive which produces small air bubbles. The air occupies the place of the coarse aggregate. The density of the concrete is generally less than 1,2 g/cm^3 (75 lb/ft^3)."

If the cell-like pores are caused by gas-producing substances, the concrete is called a gas concrete. If, however, foaming agents are added to the liquid concrete and the mix is then foamed by physical means, the result is a foamed concrete. The concretes may be hardened either under normal ambient conditions or else heat treated or hardened under steam pressure. Since only low strengths are obtained by normal hardening processes and in addition this results in high shrinkage, high quality porous concretes are normally steam hardened under pressure at about 175 °C ($\approx$ 8 atmospheres) to 190 °C ($\approx$ 12 atmospheres).

4.6.2.1. Heat-resistant gas concrete

Gas concrete, like all the porous concretes, possesses good insulation properties. At temperatures above 600 °C, however, normal gas concrete breaks down as a result of thermal stresses and the presence of free lime in the cement binder. Basic research undertaken by *G. A. Sadovnikov* [119], has, however, shown that it is possible to make a special heat-resistant gas concrete based upon portland cement.

The gas-producing agent used is aluminium powder, with a specific surface of 600 cm^2/cm^3 (236 in^2/in^3). Aluminium paste is also suitable for this purpose. The powder is annealed at 200 °C before use, since the particles are covered with a film-like coating of 1 to 2% of mineral oil, on account of the danger of explosion and dust (the percentage is referred to the powder). The filler material consists of finely ground chamotte. Chamotte powder of a particle size up to 0,06 mm (0,0024 in) in the proportion of 12% (referred to the weight of the cement) may also be added as a ceramic stabiliser. The functions are the same as in all other concretes bound with portland cement. Fly ashes from thermal power stations have not proved successful as filler materials. When the concrete is heated above 600 °C the surface develops a network of fine hair-cracks, which may partly damage the structure. These test-pieces normally lose 20% of their cold strength. Portland cement is used as the binding agent. The optimum proportions of the mix are given in Table 10. The densities achieved lie between 600 and 800 kg/m³ (37,3 and 49,7 lb/ft³).

Table 10. Composition of heat-resistant gas concretes (Sadovnikov)

Constituents of mix	Proportions [kg/m³ concrete]	
	a	b
Portland cement	250	250
Finely ground chamotte	125	250
Chamotte powder < 0,6 mm	30	—
Gas-producing agent PAK-3	1,5	1,5
Water/cement ratio	0,7	0,88

This heat-resistant gas concrete is made in a mortar mixer, following the normal procedure for gas concretes, the filler material being previously mixed with the cement in a dry state. The water is pre-heated to 50 °C and added last of all. The mixing process is continued until a homogeneous, pourable, mix results: the easier the mix is to pour, the more it will be expanded by the subsequent evolution of the gas. The aluminium is then added and the whole is mixed a second time for a period long enough to allow the gas formation to take place. Finally the concrete is poured into the forms, filling them to about two-thirds of their height. The expansion of the concrete continues until the cement sets, filling the forms completely.

The concrete is left in the forms until the necessary strength is reached and is then cut into blocks or slabs of the required size. These elements are then heated in an autoclave or by electrical means. As a result of the use of hot water in making the concrete, the dwell time necessary in the forms is reduced. The high temperature of the elements at the start of their heat treatment has a good effect in that it reduces the temperature stresses, thus preventing the formation of cracks. If this were not done the autoclave treatment could result in extensive cracking.

The following hardening procedure has been proved by a number of tests at a pressure of 8 atmospheres; raising of steam to pressure — 4 hours; pressure held constant at 8 atmospheres — 4 hours; cooling — 4 hours.

Heat-resistant gas concretes are also manufactured using normal hardening procedures, on account of the expense of the autoclave treatment and the necessary mass production plant associated with it. The shrinkage resulting from the drying out process is always a drawback of this method. In addition the strengths achieved are usually lower. The following mix proportions were developed and tested in the Peoples' Republic of Poland, the quantities stated being for 1 cubic metre of concrete:

300 to 360 kg finely ground chamotte (504 to 604 lb/yd³),
320 to 360 kg portland cement, quality 350 or 400 (536 to 604 lb/yd³),
0,33 kg aluminium powder (0,55 lb/yd³),
10 kg gypsum (16,8 lb/yd³),
7 kg Na_2SO_3 (11,8 lb/yd³,
330 to 345 kg water (544 to 580 lb/yd³).

The gypsum is added principally in the form of a semihydrate ($CaSO_4 \cdot 0,5H_2O$) for the purpose of controlling the setting time of the concrete.

The elements are cast in forms measuring 800 × 800 × 100 mm (31,5 × 31,5 × 4 in). Because of the sensitivity of gas concrete to early drying out, if it has not been hardened under steam pressure, and also in order to obtain satisfactory strength, these concretes should be maintained in a thoroughly wet condition for at least 7 days.

4.6.2.2. Heat-resistant foamed concrete

The maximum temperature at which normal foamed concretes can be operated, subject to special precautions being taken, is 500 °C; this has been demonstrated by *I. T. Kudrjašev* and *V. P. Kuprjanov* [145]. It is possible to improve the resistance so that higher temperatures can be tolerated, but only if very fine ceramic aggregates are added. *K. Martin* reports success in producing a heat-resisting foamed concrete [118] having a density of 1,1 to 1,2 (68,5 to 74,8 lb/ft³) and capable of operating at temperatures up to 1100 or 1150 °C; this concrete was hardened normally, without any subsequent treatment in an autoclave. Curing under warm conditions in any case shows very few advantages with foamed concrete. It produces an early gain of strength only if blast furnace cement is used; this is sometimes done to enable the precast units to be removed and stacked elsewhere. Steam curing does not lead to any shortening of the shrinkage time nor does it decrease the total amount of shrinkage. If portland cement is used the drying out and consequently the dimensional change of a foamed concrete may even be delayed [144].

Water/cement ratios vary between 0,6 and 0,9. The required proportion of foaming agent is about 0,8 to 1,0% of the weight of the cement. The cement should be in a proportion of from 30 to 50% if satisfactory strength is to be achieved.

The manufacture of foamed concrete on a commercial scale is carried out as follows:

The foaming agent is worked up into a stable foam; the binding agent and fine aggregates are mixed into a plastic state as a separate operation. Finally the latter mixture is combined with the foam and mixed into a foamed wet concrete. Horizontal-shafted paddle mixers are used for agitating the foam and mixing the mortar, and a double-shafted pan mixer with horizontal paddles for mixing the foamed concrete. Two systems are available for making the initial mixture of foam; first the air-entraining method, in which the solution of foaming agent is beaten up by means of a mechanical agitator and secondly a system in which the solution is pneumatically agitated. The latter method is to be preferred because it produces a more stable foam and also because it requires a smaller quantity of foaming agent. The wet concrete can either be cast into the forms directly or, if required, pumped into a number of forms stacked one above another.

Since all foaming agents lead to a slowing down of the setting process, it is not possible to remove the shuttering at an early stage even when smelted aluminous cement is used. *K. Martin* [118] states a standing time in the forms of 15 hours for the mixes upon which he carried out experiments. The cold compressive strength after 7 days was between 35 and 70 kp/cm^2 and the compressive strength after heating to 1000 °C between 14 and 15 kp/cm^2 (500 to 1000 psi and 200 to 213 psi respectively).

Finally reference should be made to the possibility of making a foamed concrete for high temperatures using lightweight aggregates. In the standard TGL 117-0377 [146] this is understood to mean a porous concrete made for normal temperatures using a foamed mortar and lightweight aggregates, which is then hardened by hydrothermal treatment. The particle size of the lightweight aggregates is given, for the case of blast furnace pumice, as 5 to 25 mm (0,2 to 1 in). Chamotte- or brick-dust had to be used for the very fine fraction of the aggregate. Concretes of this type require very little cement and shrink to a less extent than normal foamed concretes. The maximum shrinkage permitted after hardening is 0,5 mm per metre (0,05%).

This concrete is mixed in a pan-mixer. In the first place the fine aggregate and the binding agent are combined with the required quantity of water and the foaming agent and the whole is mixed and agitated for 2 to 3 minutes. When the density specified for the foamed mortar has been reached, the lightweight aggregate is added and the whole is then mixed again. The total mixing time including foaming time should be 5 to 6 minutes. The final hydrothermal treatment is carried out by means of low pressure steam, the forms being removed for this purpose. According to the standard referred to above the following steam-curing schedule should be used:

        1 hour up to 30 °C,
        1 hour to raise from 30 to 80 °C,
        7 hours held at a minimum of 80 °C,
        1 hour for cooling down to 20 °C.

It should be possible to develop concretes of this type for high temperature use as well and to utilise them effectively.

5. Properties of Refractory Concretes

A high temperature concrete in its hardened state (in the nomenclature of general concrete technology, a set, as opposed to a wet, concrete) must satisfy certain specific requirements which arise from its intended conditions of operation in the furnace. These may be either laid down in the relevant Standards or may have been specified as a result of previous operating experience. On the other hand the refractory concrete itself possesses certain properties; it is important to be aware of these and to take them into account in choosing the correct materials and in reaching a proper understanding of its behaviour under working conditions. Basically, there are certain properties, as with all refractory materials, that come into play and have their effect at high operating temperatures and that, therefore, in the last resort either determine the possible use of the material in refractory plant or else must be considered in the design of the concrete. These more important properties include the following:

Refractoriness, softening under pressure and resistance to creep, elasticity and creep behaviour, thermal conductivity, porosity and permeability, thermal expansion, shrinkage and after-shrinkage (or second-stage shrinkage), stability under variable temperatures, stability in the presence of chemicals and slags.

There are numerous differences, in thermal behaviour and in many other properties, between the heat-resistant and fire-resistant concretes made with cement and those that are made using other binding agents of an anhydraulic, inorganic nature which set without the use of heat. It is therefore advisable to treat the different types of concrete separately. There are therefore separate sections below for each of the different types of high temperature concrete, classified according to the type of binding agent used, such as cement, waterglass, magnesia and phosphate. It will be evident that it is possible to discuss only a limited number of their more important features. The available technical literature offers to the reader a wealth of detailed information which it is not intended to repeat here, even were it possible. In case of the need for specialist information, therefore, reference should be made to the relevant original work quoted in the list of references. It must be noted in this connection, however, that many of the technical properties quoted here were obtained under specific circumstances and cannot therefore necessarily be used in every context as if they were universally applicable.

5.1.   Cement-bonded Concretes for High Temperatures

The most outstanding characteristic of the cement-based refractory concretes, when contrasted with all the other numerous refractory materials, is that many of the properties listed above exhibit quite different values during the operating life of the concrete in its refractory function, from those obtained at room temperature or during the first excursion to high temperature. The reason is that many chemical changes take place under the influence of heat, as already seen when discussing the cements, thereby altering fundamentally the character of the original concrete. It is transformed from a typical concrete, that is to say, a heterogeneous bond of aggregates and binding agent, into a ceramically hardened structure due to the high temperature it has undergone and the associated smelting and sintering reactions.

The intermediate and final phases which result from these reactions are particularly characteristic and of particular importance for cement-bonded refractory concrete. In order to arrive at a proper understanding of its behaviour, therefore, it is essential that the refractory engineer study the pyrochemical transformations that take place in the concrete before he considers the resultant properties which are his main concern.

5.1.1.   Pyrochemical Processes

The term pyrochemical process is used in the text below to describe all those processes, reactions, changes and transformations that occur in industrial refractory concretes, usually upon first heating but sometimes on subsequent reheating. Consideration has already been given, in Sections 2.1.2 and 2.1.4, to the chemical alterations that occur in pure portland and aluminous cements after they have set, as a result of temperature. The following discussion relates, in the first place, to the chemical aspects of the processes of interaction between the intermediate phases of the cements and the aggregates together with the phase-building which results therefrom. The effects of these changes upon the important properties of the concrete, for example, the strength, expansion, shrinkage and stability under variable temperature, are dealt with in Section 5.1.2.

5.1.1.1.   Refractory concretes made with portland cement and slag cement

As already stated, portland cement differs fundamentally from aluminous cement in its pyrochemical behaviour, while the slag cements occupy a place somewhere between the two. One of the most decisive factors is that free lime arises in portland cement when it is heated. Under certain circumstances this may determine its properties and limit its use and, therefore, as has already been pointed out, certain steps must be taken in the manufacture of concretes made from portland cement.

5.1.1.1.1. The Problem of Free Lime

The following is a recapitulation of the processes that take place during the hardening and subsequent heating at a high temperature of a pure, set, cement.

The hydration of the highly basic silicates which constitute the most important clinker minerals results in the formation of tobermorite-like phases and the release of large quantities of $Ca(OH)_2$. In the case of slag and pozzolanic cements it is implicit that these quantities are fairly small, due to the low content of portland cement clinker and the combination of the free lime with slag or pozzolana. When the temperature is raised to 500 °C, the calcium hydrate loses its combined water and changes into very reactive CaO. It is only to be expected that when this CaO cools to a low temperature, as is often the case in the

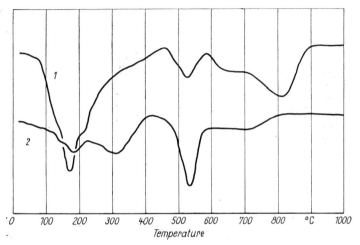

Fig. 35. DTA-curves for hydrated portland cement (above) and hydrated portland cement after heating to 600 °C and storing under moist-conditions (below)

1 Hydrated portland cement, unheated; 2 hydrated portland cement, heated to 600 °C and stored moist

operational history of a furnace, it undergoes a reaction with the moisture in the air which results in the formation of $Ca(OH)_2$ once again, followed by "blowing" of the cement and consequently by destruction of the structure of the set paste. It is well known that the density of CaO is 3,37 (210 lb/ft³), while that of $Ca(OH)_2$ is 2,34 g/cm³ (146 lb/ft³); the former therefore expands upon hydration by approximately 44%. In earlier literature this fact was always used as a decisive argument against the use of portland cement in refractory concretes. This argument is sometimes still found in modern works.

In principle this argument is not wrong. Several examples are quoted by *K. D. Nekrasov* [6; pages 24, 51]; hydrated alit completely loses its strength after burning at 1200 °C followed by storage in dry air, while a set portland cement shows signs of disintegrating if stored for a period after being heated to a temperature in the range 600 to 1000 °C. Similar observations have been reported by *C. M. Vogrin* and *H. Heep* [147] and *A. F. Livovich* [148]. The authors' own investigations have demonstrated the following: after heating the hardened cement to 600 °C and storing it for a week under moist conditions, the differential thermo-analysis indicates a marked $Ca(OH)_2$-effect (see Fig. 35). The evidence of disintegration may be seen from the test cubes in Fig. 36.

It was some time after *Platzmann* had stated in a patent [19] that it was possible to combine the free lime with either trass or free silica, that Soviet scientists followed up this long forgotten discovery and produced a fully developed method of utilising portland cement in refractory concretes, by binding or combining the lime which is released during the hydration process and so rendering it innocuous. It is already known from the technology of cements

Fig. 36. Damage as a result of the re-hydration of free CaO
Left: a portland cement block which has been heated to 900 °C and then stored in water. *Centre and right:* blocks made from portland cement and chamotte powder and from portland cement, chamotte powder and clay dust, after similar treatment

that this can be done by adding, for example, slags in the form of slag cements, or pozzolana (trass-portland cement). These additives combine with the $Ca(OH)_2$ while the process of hydration is still going on, resulting in the formation of calcium silicate and aluminate hydrates. Another possibility is to cause the lime to combine at high temperatures, that is while being first raised to temperature or during operation; in this case the lime is combined with very fine-grained substances which are acid or of low lime content, during the course of pyrochemical reactions. This can be achieved, as before, by the use of slags or pozzolanas, but in refractory concrete technology the substances used are generally the very fine-grained, fire-resistant "ceramic stabilisers" the more important of which have already been discussed in Section 2.4. Reactions of this type are inevitably accompanied by volumetric increase of the material [147], but so far any blowing effects that may have resulted have not been observed.

It has been found from extensive investigations into the mode of operation of the very fine aggregates and their effectiveness, that the combining of the free lime is dependent upon the temperature, the type of ceramic stabiliser used and also upon the quantity of the latter. Really effective reactions do not usually start before a temperature of at least 600 °C is reached. Soviet research has shown that the proportion of the filler aggregates may vary over a very wide range, from 30 to 100% of the cement content. In certain cases, for example when magnesia or chromite are used, the quantity may be still higher. In another reference [103] 25 to 30% is stated to be a satisfactory proportion. A final figure can be established only by experiment and will depend upon the type of cement used and the mineral composition and reactivity of the stabiliser. Some typical curves showing the variation of the content of free lime in a hardened portland cement structure with temperature and type of very fine aggregate used, are indicated in Fig. 37. These have been drawn from

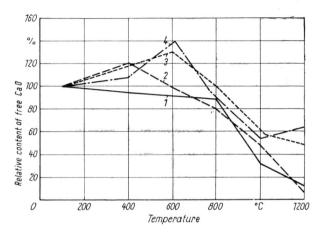

Fig. 37. The variation of the free lime content of portland cement with the temperature of preheat and the type of ceramic stabiliser used

The free CaO-content of the unheated sample is taken as 100%.
1 Chamotte dust; *2* Quartz dust; *3* Waste slag;
4 Granulated blast-furnace slag

the works of *K. D. Nekrasov* and his colleagues and refer to a cement/aggregate ratio of 70:30. From these curves it can be seen that chamotte dust and quartz dust are better than the slags. According to *L. Ludera* [103] the best stabilisers are fire-resistant clays, chrome ore, chamotte dust and brick dust; silica dust and flue-dust are rather less effective. The effectiveness of sintered magnesia arises from the reactivity of the constituents, SiO_2, Al_2O_3, and Fe_2O_3; their total effect is, however, relatively small but can be augmented considerably by the simultaneous addition of chrome ore (the proportion stated in the U.S.S.R. references being about 30% of the weight of cement).

In the case of slag cements, the proportion of clinker and thus the amount of free lime that occurs is less, due to the presence of the slag. Furthermore, since the slag is able to combine with the lime, the use of very fine filler aggregates may be found unnecessary, if certain conditions are fulfilled; for example, if the operating temperature is moderate and if the slag content is at least 50% [8].

Depending upon the particular operating temperature, the following materials may be used as ceramic stabilisers [103]:

up to 1250 °C — chrome ore,
up to 1100 °C — chamotte dust,
up to 800 °C — brick dust.

It must be mentioned that the very fine additives have other functions as ceramic stabilisers, quite apart from that of combining with the lime; these are discussed in more detail below.

5.1.1.1.2. Formation of New Phases

When refractory concretes are heated a number of reactions take place, which are characterised on the one hand by the decomposition of the hardened cement structure, together with the associated formation of new phases, and on the other hand by the reaction of these intermediate phases with the additives and aggregates. Among the decisive influences on the properties of refractory concrete are the development of various new minerals that occur during the last-named processes due to chemical combinations with the coarse fire-resistant aggregates.

The decomposition of the hardened cement structure occurs between 200 and 800 °C. It commences with the driving off of the water that lies in the gel and between the layers of the tobermorite-like phases and the decomposition of the ettringite, continues with the dehydration of the $Ca(OH)_2$, and ends with the transformation of the dessicated hydrosilicate into stable water-free compounds, such as CS and C_2S. At very high temperatures (≈ 1300 °C), the free lime content again decreases somewhat, since it undergoes reactions; for example, with CS to form C_2S. These high temperature reactions never take place in isolation however, but (in concrete) always in association with reactions between the constituents of the cement and the aggregates. Reactions of this type begin in the solid phase, before any smelted material is present, at temperatures which have been estimated to lie between 700 and 800 °C.

The results of thermograms (thermogravimetric analyses and, more especially, differential thermo-analyses) are a qualitative guide to the occurrence of reactions in refractory concretes. Under certain circumstances they may to some extent be quantitative as well. The DTA curves given in Fig. 38, which have been compiled from various technical literature and from the authors' own experiments, are for hydrated mixtures of pure clinker mineral with finely ground chamotte and for portland cement with chamotte dust. They show the general character of the unmixed constituents and contain few special effects. Apart from the endothermic effects already observed (see Figs. 3 and 4) the only other important feature is an exothermic effect at about 1000 or 1100 °C, which can be attributed to the formation of anorthite, CAS_2 and rankinite, C_3S_2. All these effects are only clearly defined in the case of mixes with fairly high cement contents; they become obscured in lean concretes.

K. D. Nekrasov [6] has reported that microscopic examination of samples that have been subjected to high temperature also indicates that anorthite and rankinite are formed. Mullite is also sometimes found among the newly formed phases. Among the mineral substances that separate out from the pure cement matrix is wollastonite; the quantity of this is considerably increased if stabilisers containing quartz are added or if crystallisation of slag constituents takes place. There is, however, always a certain amount of general vitreous material formed, in quantities which will depend upon the separate constituents present.

The results of radiographic investigations by *H. Gibbels* [55] have shown that after annealing heat-resistant concretes made with blast-furnace cement and chamotte at temperatures in the range 900 to 1100 °C a lime-felspar phase was formed.

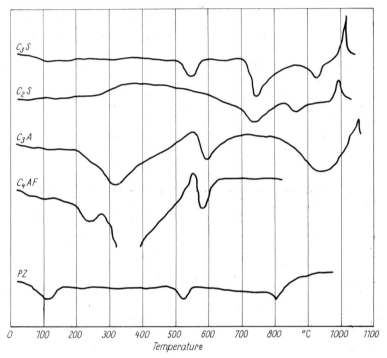

Fig. 38. DTA curves for hydrated clinker minerals and portland cement with the addition of chamotte powder
PZ = Portland cement

There is very little information in published literature about the formation of minerals in portland cement refractory concretes which contain additives of magnesia, chromite, and so on. It is known from investigations by *G. D. Salmanov* [149] that monocalcium chromite, $CaO \cdot Cr_2O_3$, which is fire-resistant up to 1850 °C, is formed from the chromic oxide of the chrome ore and the free lime of the cement. This reaction starts at a fairly early stage; CaO and Cr_2O_3 react exothermally from 720 °C upwards [150]. Magnesia reacts hardly at all with the constituents of the cement; its main function is to form highly temperature-resistant compounds with the easily smelted products of the chromite.

There is one phenomenon which should be particularly noted in connection with the phase-building processes that occur in refractory concretes made with portland cement. Among the compounds that are formed by the reaction of the free lime with the dessicated CSH-phases as a result of heating is C_2S. C_2S can also occur due to admixtures of sintered magnesia in the concrete. If the concrete is heated to high temperatures and then cooled down, C_2S may change from the β-form to the γ-form below 600 °C, resulting in a volumetric increase which can destroy the structure of the concrete. Soviet sources state that in this case the addition of phosphorite powder ($Ca_3(PO_4)_2$) (1,2% P_2O_5 referred to the cement) can be of assistance [6], or the addition of boric acid to the cement or to the chamotte [151]. A boron content in the clinker of 0,5 to 3% is mentioned, or of 1 to 7% in the chamotte (calculated on the elementary boron). The clinker and the chamotte are then ground up together [152][1]. The phenomenon of the destruction of the so-called clinker concrete lining, which at one time was used in rotary cement kilns (see also Section 6.6), as a result of the β-γ-C_2S-transformation, was combated by *V. N. Jung* and *A. I. Koršunova* [153], [154] in the early 1930s by the use of boric acid or chromic oxide additives.

The formation of mineral phases in slag cements follows along lines similar to the processes described for pure portland cements. In addition, crystallisation of the slag takes place; this begins to be clearly observable at 1000 °C.

5.1.1.2. Refractory concretes made with aluminous cements

In addition to the higher thermal resistance offered by aluminous cements, particularly certain special types, the special advantages of aluminous cements in the making of fire-resistant concretes have always been acknowledged to be the way in which they lose their water gradually when heated and the absence of free lime in the sintered product. These qualities favour a stable concrete structure. Now that the problem of free lime has been overcome in portland cement concretes by the use of ceramic stabilisers, the principal advantage that the aluminous cements still retain is the gradual way in which they give up their water. The manner in which the water is driven off is in fact relatively simple, there being far less discontinuity in the process than in the case of the silicate cements.

A comparison of the DTA-curve for pure aluminous cement (Fig. 12) with that for a smelted aluminous cement with chamotte aggregate, which is given in Fig. 39, shows virtually no differences in principle. It is evident that the effects are less marked and that, amongst other things, they are shifted to a lower temperature range. A slight exopeak (exothermal peak) may sometimes occur at about 1000 °C. This may possibly be attributable to a reaction of the cement with the chamotte and will of course be unlikely to manifest itself if less active or inactive materials, such as corundum, are used.

H. Mitusch [12] has carried out detailed and fundamental research into the phase-building processes that occur in concretes made with smelted aluminous cements and different types of aggregate. This investigator has studied the condition of refractory concretes having a ratio of smelted aluminous cement

---

[1]) The introduction of boric acid to the cement and the aggregate can raise the residual strength of the concrete in the critical temperature region in certain circumstances. This is not true, however, for all concrete mixes [6].

to aggregate of 1:1 after five hours of firing at temperatures up to 1350 °C. At about 500 °C γ-alumina started to appear, as an intermediate phase. Some gehlenite, both newly formed and original, and some non-hydrated β-C_2S, were also present. In chamotte concrete CA_2 appeared at about 1000 °C, but this decreased in quantity due to subsequent reactions. Finally, anorthite, CAS_2, and gehlenite, C_2AS, appeared as new phases at 1350 °C. A most remarkable fact was that the chamotte minerals could then no longer be traced; they had therefore been completely transformed. This is only true, however, for the mix

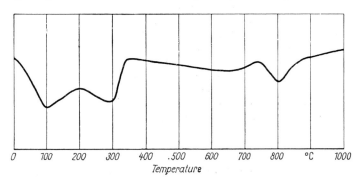

Fig. 39. DTA-curve for smelted aluminous cement with chamotte aggregate

proportions stated above, since with different proportions of smelted aluminous cement to chamotte a different distribution of phases could occur. The phase diagram indicates that in all cases anorthite and gehlenite will exist as final products.

If sintered magnesia is employed as the aggregate CA_2 is absent, because reactions will have taken place earlier with the intermediate mass (the matrix between the crystals) of the sintered magnesia[1]. At 1000 °C gehlenite, β-C_2S, and reformed $C_{12}A_7$ are to be found as well as periclase. The gehlenite decreases at higher temperatures, spinell MA appearing instead.

The many and varied phase-changes and phase-forming processes that take place during the heating of refractory concretes, together with the vitreous phases that arise as well, lead to the development of ceramic bonding and so to a second stage solidification in the structure of the material. A pre-requisite for this is an adequately high temperature, which will enable the transformations to develop fully and ensure the cementing and the sintering of the highly viscous amorphous phases. The aggregates naturally play an important part in these processes; American workers have stressed the importance of the effects obtained by the introduction of easily vitrified clays to assist in the development of ceramic strength [38], [105].

There is one problem associated with the first heating of refractory concretes that should be especially noted; this is the tendency to fracture in a manner that is almost explosive (spalling). This process, which can lead to sudden

---

[1] According to *H. Mitusch*, only the intermediate mass, which is of course composed of the flux, reacts in these circumstances; periclase appears to behave initially in an inert manner.

destruction of the concrete, is affected by a number of factors. *W. H. Gitzen* and *L. D. Hart* [155] have listed the following; permeability, strength, thermal conductivity, aggregate/cement ratio, water/cement ratio, the granulometry of the aggregates, the mixing procedure, the dimensions of the unit of concrete, and the rate of increase of temperature. Spalling is brought on by the sudden development of moisture vapour, arising from the dehydration of various hydrated compounds.

W. H. Gitzen and *L. D. Hart* have paid particular attention to this phenomenon [155] and have studied the conditions that are necessary to eliminate or reduce spalling. A procedure which has proved very effective is to subject the concrete to one day of heat-treatment at temperatures in the range 20 to 32 °C followed by drying at 105 °C. This method may be used for concretes made either with normal aluminous cement or with high alumina cements. A similar result may be achieved either by increasing the particle size of the cement and thus increasing its permeability or by reducing the cement content and the water/cement ratio.

It should also be mentioned that *D. F. King* and *A. L. Renkey* [156] recommend the use of boric acid as an additive, for preventing explosive spalling.

5.1.2.   Irreversible Temperature-induced Changes in the Physical- -mechanical Properties of Concretes

The reactions of decomposition and phase-building which result from the heating of high temperature concretes lead to a number of other changes to the properties of the concretes. These changes are dependent upon the type of heat-treatment to which the concrete has been subjected and involve an intermediate stage characterised by its own specific material properties, before the final stage is reached. In the final stage the concrete is subjected to the highest temperature and undergoes a long dwell period at its normal operating temperature, which causes it to reach a condition in which its properties are reproducible, being affected either very little or not at all by subsequent repeated fluctuations of temperature. It is important to have a proper understanding of variations in properties that are temperature-dependent in this way, in order to assess the probable effects upon the concrete of the conditions to which it will be subjected in service. Only by this means can shortcomings in design be avoided.

5.1.2.1.   Cold compressive strength

As has been seen in Sections 2.1.2 and 2.1.4 and in Figs. 5, 12 and 13, cements, when subjected to heat, usually pass through a stage of minimum strength. The temperature at which this occurs varies considerably, depending upon the type of cement. Similar phenomena evidently occur in concretes, as their strength when heated is primarily determined by the strength of the binding agent. While the strength of the fire-resistant aggregate varies little or not at all with temperature (except of course at very high temperatures, which normally cause all fire-resistant materials to lose strength) the strength of the hardened cement structure varies as a result of changes to that structure. This cement structure is therefore the weakest member in the system, in the range which lies between the phases of hydraulic and of ceramic binding, and is therefore critical to the whole strength behaviour of the material.

The behaviour of refractory concretes is usually assessed in terms of the temperature-dependent changes of strength (usually the compressive strength is considered) which arise during pre-heating. A simplified method of making this assessment is by measurement of the cold compressive strength after the concrete has been subjected to specified thermal conditions. The reader who is new to the subject may well ask whether this gives a true picture of the real state of affairs. It is well known that the cold compressive strength of a material which has been heated to a particular temperature does not always correspond to the strength at that temperature; the maxima exhibited by chamotte and silica at higher temperatures will be remembered as an example. Nevertheless, it has been pointed out by *Nekrasov* that the hot compressive strengths in the region of 800 to 1000 °C of concretes made from portland cement are of the same order as the cold strengths obtained from tests after the concrete has been cooled down to room temperature. It can therefore be assumed that the cold compressive strength, which is much more easy to determine, can give a useful measure of the real conditions, or at least of their tendency to vary.

5.1.2.1.1. High-temperature Concretes made with Portland Cement

The cold compressive strength of portland cement-based refractory concretes is dependent upon many factors. Among the more important are the type of binding agent, the cement/aggregate ratio, the water/cement ratio, the type, shape and grading of the aggregates, the resultant density of packing, the standing time of the wet concrete and the reactivity between the binding agent and the very fine aggregate. The strength will, however, be decisively influenced by the thermal behaviour of the binding agent itself. The clinker minerals show wide variations in this respect. For example, the very fine aggregates function much more effectively in conjunction with hydrated alit and C_3A than with hydrated belit.

It is characteristic of high temperature concretes of this type, as it is of the pure cements, that there is an initial increase of strength up to about 300 °C, followed by a decrease in the middle temperature range. A typical curve of compressive strength related to temperature of pre-heating, for a portland cement concrete, is given in Fig. 40 (*L. Ludera* [103]). The minimum strength is in the range 600 to 1000 °C. The decrease is very variable, the loss of strength being from 20 to 50% of the original value. Ceramic stabilisers have a very good influence in part upon the strengths at lower temperatures; not only do they prevent a sharp loss of strength, but they also usually increase the strength in the range 200 to 300 °C above the initial strength figure.

The type of fine aggregate employed can have a significant influence upon the strength. Chamotte is normally found to possess the best properties in this respect. Clay dust acts in a similar manner. If the correct proportion of very fine aggregate is used it should be possible to limit the loss of strength to an almost negligible amount. (*Nekrasov*, for example, recommends that a proportion of more than 70% by weight of the binding agent should be used.)

Quartz dust and ground slag are not so effective, although even these materials give a noticeable improvement in the strength. A material which is still less effective is fly-ash; in certain circumstances, however, it can be a useful stabiliser.

The optimum quantities of materials such as these will vary according to the type of material and the type of cement. The results of a large number of laboratory tests have shown that the ratio of binding agent:aggregate can vary through the range from 1:0,5 to 1:3, and still give satisfactory results.

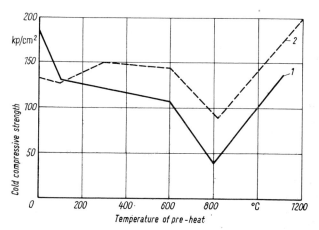

Fig. 40.   Cold compressive strength, after pre-heating, of portland cement concrete containing aggregates of chamotte and chamotte + refractory clay (Ludera)
1 Chamotte aggregate; 2 Chamotte + refractory clay aggregate

The best results are always found, however, when the quantity of chamotte or other very fine aggregate is larger than that of cement. The lowest possible limit of chrome ore + magnesia is given as 0,2 of the cement content. If a smaller quantity is used it is not possible to ensure that the free lime is all fully combined. The relationship between the cold compressive strength and the proportion of chamotte should be found by experiment; the highest practicable proportion of stabiliser gives the most satisfactory results.

The fineness of the particles of the ceramic stabiliser is no less important in its effect upon the strength behaviour. Systematic investigations have shown that the finer the aggregate, the higher will be the cold compressive strength before heating takes place and the smaller will be the relative loss of strength in the temperature range 600 to 1000 °C.

In relation to the water/cement ratio it may be stated that the cold compressive strength of a refractory concrete which has been subjected to heat is influenced by this ratio in the same manner as the strength of other concretes; the higher the water/cement ratio, the lower the values of the absolute strengths and the greater the relative loss of strength.

Another factor which affects the strengths of refractory concretes after heating is the time that elapses before the concrete is first subjected to heat. The authors are indebted to *I. E. Gurvič* for the results of his relevant investigations with portland cement and portland cement concretes [29]. The cold compressive strength after heating increases with age in the same way as the normal strength. Thorough hydration of the clinker minerals is therefore conducive to good chemical reactivity. This is understandable when it is remembered that, as a result of the formation of gels and very finely distributed mineral phases of low lime content during the hydration process, in addition to the formation of hydrate of lime, substances are produced which are more reactive than the clinker minerals themselves. The coating of the particles of additive and aggregates and the consequent reactions that take place with them is naturally better in the case of thorough hydration than in a partly hydrated state. The influence of ageing is only significant, however, at low temperatures, where it affects the hydrated strength; it is much less at high temperatures.

It is important in relation to practical concreting to know whether hardening accelerators, such as calcium chloride, affect the thermal and mechanical properties of the concrete after it has been fired. Investigations have shown that $CaCl_2$ is not very suitable on either count. The strengths are usually lower than in concretes made without the use of accelerators, Soviet experience having demonstrated that $CaCl_2$ is suitable for use only in normal concretes for temperatures up to 600 °C [6; p. 126]. *L. A. Cejtlin* [110] has been able to demonstrate that in the case of concretes made with pozzolanic cements, the addition of $CaCl_2$ raises the strength at 600 °C. In the range 800 to 1000 °C, however, it falls off again and becomes the same as that of concretes made without accelerators. A proportion of 2% of $CaCl_2$ has been used with satisfactory results in chamotte concrete made with pozzolanic cement, for use as insulation in pre-heating furnaces for metallurgical processes [157].

A remarkable phenomenon has been observed when carrying out tests to ascertain the cold compressive strength of concretes that have been subjected to heat. There is a difference between the strength of cubes tested immediately after heating and those that are not tested until they have been stored for a period in moist air; the latter always showed the higher strength. When therefore the combination of the free lime is complete (and this is certainly so in concretes that contain adequate stabilisers) there is still a favourable effect to be obtained from the influence of moisture upon the concrete after it has cooled down.

An improvement in concrete strength can also be obtained by hot-curing, as is well known from the experience of general concrete technology. *A. E. Fedorov* [158] has shown that this is also true for heat- and fire-resistant concretes made with portland cement and chamotte. It can be seen from Fig. 41 that autoclave treatment gives better results in this respect than simple steam-curing. For all three methods of hardening shown, however, the relative strengths over the temperature range are approximately the same. It was also found possible to obtain similar good results, under certain special conditions, by means of electrical treatment during the hardening process [159].

In the past it has been the cold compressive strength that has been the subject of tests. The same is likely to be the case in the future. Tensile or bending strength tests are usually ignored. It is interesting to note, in this connection, that the tensile strength behaves in a manner rather different from the compressive strength. This has been demonstrated by *P. P. Budnikov* and *D. S. Il'in* [33]. While the compressive strength usually increases at first, in a continuous manner, and begins to decrease after 300 or 400 °C is reached, not falling to a value below that of the unheated concrete until the temperature is about 600 °C, the tensile strength nearly always drops as soon as heating commences and is very low indeed at 500 to 600 °C. Similar results have been

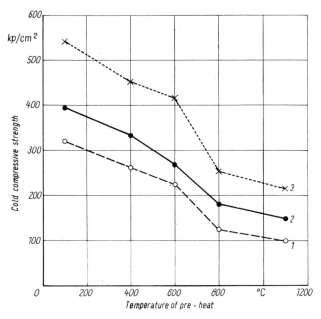

Fig. 41. *Variation of the cold compressive strength of portland cement/chamotte concrete with temperature of pre-heat and method of curing*

1 Stored wet for 7 days; *2* Steam cured for 12 hr at 85 °C; *3* Cured in autoclave for 12 hr at 170 °C

reported by *K. D. Nekrasov* [6]. The tensile strength falls continuously up to temperatures above 1100 °C. Its absolute value is also naturally very much dependent upon the type of cement used and the fine additives. This behaviour would appear to be very important in any assessment of the adequacy of refractory concretes.

5.1.2.1.2. Refractory Concretes made with Slag and Pozzolanic Cements

Refractory concretes of this type behave in principle in a manner similar to the portland cement concretes, but their strengths are usually somewhat lower. The minimum value of the strength will be more or less marked, depending upon the content of slag or pozzolana. It is often hardly distinguishable if the right proportions are chosen. In the case of slag cements the increase of strength is usually displaced to a lower temperature, on account of the earlier sintering of the slag constituent.

Some curves of cold compressive strength, as published by *H. Gibbels* [55] and *I. Schmid* [104] for refractory concretes made from iron portland cement and blast-furnace cement, are given in Fig. 42. Certain differences can be recognised here, which are attributable to the particular types of cement; caution must therefore be exercised in the use of these curves for detailed

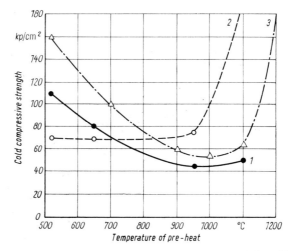

Fig. 42. *Variation with temperature of pre-heat of the cold compressive strength of refractory concretes made with iron portland cement and blast-furnace cement (Gibbels, Schmid)*

1 Iron portland cement + chamotte; *2* Blast-furnace cement + chamotte; *3* Blast-furnace cement + chamotte + clay

values. The remarkably small loss of strength in the middle temperature range which can be seen in some cases, is attributable to the fact that in the slag cements the finely ground component of the slag acts as a ceramic stabiliser. The pozzolanas in pozzolanic cements act in a similar manner.

5.1.2.1.3. Fire-resistant Concretes made with Aluminous Cements

There has been much observation and research on the behaviour of the strength of aluminous cement concretes in relation to the temperature of preheating. This knowledge goes right back to the first utilisation of aluminous cement about 35 years ago. *H. Mitusch* [12] has undertaken very detailed and fundamental research into the various influences that can affect this property, in recent years.

It is a characteristic of fire-resistant concretes made with aluminous cements, that the fall-off of strength takes place over a very wide range of temperature. Fig. 43 shows a number of typical curves selected from the relevant literature.

In contrast to portland cement concretes, there is no initial increase of strength with temperature; in fact there is normally an immediate decrease. In certain circumstances, however, heating to between 600 and 800 °C may result in an observable intermediate maximum value of the cold compressive strength. Such maxima have been described, in particular in the works published by *H. Mitusch* and by *J. F. Wygant* and *W. L. Bulkley* [160]. Since they are clearly dependent upon the content of smelted aluminous cement (Fig. 44), they are attributable to the formation of intermediate phases. The type of

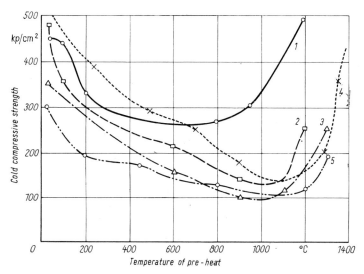

Fig. 43. Variation with temperature of pre heat of the cold compressive strength of aluminous cement concretes with various aggregates (Nekrasov, Mitusch, Kravšenko, Kuntzsch and Rabe, Heindl and Post)

1 Chrome ore; *2* Chamotte; *3* Chamotte; *4* Chamotte; *5* Corundum

aggregate also plays a part, however, since every type does not result in such clearly defined maxima. This interim increase of strength is followed in the majority of cases by a further decrease to a well-marked minimum at around 1000 to 1200 °C. This is regarded as a critical range and special attention should be paid to it when using refractory concretes made with smelted aluminous cements.

The actual shape of the curve of strength is also very much dependent upon the aluminous cements and the aggregates that are employed. Examples may be found in different technical papers of curves that are quite flat and also of ones that show very marked minima, limited to a small interval of temperature. The presence of an intermediate maximum is as common as the absence of one. The observations have also evidenced the occurrence of a very wide range of values of the loss of strength. This range may vary from a few percent to 60% or more and here again the values are very much dependent upon the type of cement and aggregate present. It is consequently impossible to lay down typical, uniform values, but there is evidence that the duration of the heating period has an influence upon the strength [161]. For instance the strength of a concrete will be higher after heating to a temperature of 900 °C for one month than after a heating period of only 5 days. It is generally typical of aluminous cement concretes, however, in contrast to portland cement concretes, that the decrease of strength is continuous over a relatively wide range of temperature and is not nearly so great.

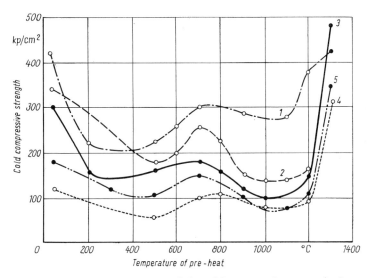

Fig. 44. Intermediate maxima of the cold compressive strength of preheated aluminous cement concretes (Mitusch)

1 Sillimanite, 20% ac; *2* Chrome ore, 20% ac; *3* Magnesia, 20% ac; *4* Chamotte, 10% ac; *5* Sillimanite, 10% ac; ac = aluminous cement

As has already been seen, the strength behaviour depends upon many factors; these include the proportions of the cement and of the aggregates, the type of aggregate, the grading, the water/cement ratio, the type and duration of curing and so on.

The higher the aggregate content of the concrete, the less in general will be the loss of strength after subjection to heat. Thus *R. A. Heindl* and *Z. A. Post* [161] and *M. F. Čebukov* (quoted by *Nekrasov*) have been able to show that a reduction of the cement content or the stage-by-stage addition of blast furnace slag lead to a widening and a flattening of the minimum on the curve. Similar tendencies can be seen in Fig. 44.

The type of the aggregate itself has a major effect both upon the absolute values of strength achieved and upon the loss of strength. Examples of materials that tend to increase the cold compressive strength are sillimanite, corundum and chromite, while chamottes containing an average amount of Al_2O_3 and sintered magnesia produce average strength figures and slags and broken brick aggregates lead to considerable reductions of strength (*Čebukov, Mitusch, Kravčenko*). In the case of the last-named aggregates the minimum is also very much flattened. *G. D. Salmanov* (quoted in [6]) found that when brick dust was added, not only was there no loss, but there was in fact on occasion, a slight gain of cold compressive strength. At the same time, tests by *Mitusch* produced very flat curves. This material is not, however, of much use in concretes made with aluminous cements and intended for high temperature purposes on account of its own low resistance to temperature. The cold compressive strengths after heating to above 1200 °C are in most cases much higher than the strengths of the same concretes in the unheated state; in other words the ceramic binding action is stronger than the hydraulic binding action.

Research by *J. F. Wygant* and *W. L. Bulkley* [160] and *H. Mitusch* [12] into the effect of the content of smelted aluminous cement upon the cold compressive strength measured after heat treatment, have shown that with increasing cement content the strengths over the whole temperature range also increase. An optimum of cement content appears to be about 40 to 50%, but above this value the proportion of aggregates becomes too small for effective development of a fire-resistant structure in the concrete. These data can vary from one case to another, however, and are very dependent upon the type of cement used, the type of aggregates and other aspects of the technology. For example, some American workers have found that a definite increase of strength occurs only up to a content of 20% of smelted aluminous cement, only a very slight further increase being observable above this value [105].

In respect of the particle size distribution of aggregates, it has been shown that, in the case also of aluminous cements, the presence of a certain amount of very fine aggregate is essential to ensure the proper development of the ceramic binding process and the formation of new phases. A figure of about 60% for the proportion of the very fine fraction (0 to 0,2 mm) is regarded as desirable [12]. A certain proportion of easily vitrified clays is often recommended for this purpose [38], [105]; this should have a good influence upon the strength, especially in the middle temperature range, but will somewhat reduce the initial strength.

H. Mitusch has shown that, in contrast to the portland cement concretes, those made with aluminous cements [12] exhibit the best values of compressive strength, both in the fired and the unfired states, if the wet concrete is worked in a plastic condition. These results relate to the specific tests carried out by that author; other sources, however, state that the best strengths are obtained if the concrete mix is stiff [160], [162]. There are no further published data available, so there is really insufficient information for a universally applicable conclusion to be drawn. There is, however, universal agreement that liquid mixes produce only low strength concretes.

The only observations available about the pre-treatment of aggregates are those of *H. Mitusch*; the use of dry aggregates has resulted in rather higher strengths, after firing at 1200 °C, than the use of pre-soaked aggregates.

An increase in the time that elapses before firing has a noticeably favourable effect upon many concretes made with smelted aluminous cement, provided that the right conditions for the hydration of aluminous cement (low temperatures) are maintained. There does not, however, appear to be any improvement in the strengths of concretes in the fired state ($T = 1250$ °C) after about 4 days. Other authors [105] have found that the only effect that the length of storage time has is upon the initial strength, the effect upon the strength in the fired condition being less definite.

5.1.2.1.4. Fire-resistant Concretes made with Special Cements

The curves of strength behaviour for fire-resistant concretes made with high alumina cement are similar in principle to those for normal aluminous cement.

Detailed information about the behaviour of barium cement is found in the work of *A. Braniski* [13]. Fig. 45 shows the relationship of cold compressive strength to temperature (unfired concrete = 100%) for barium cement concretes made with various aggregates. The proportions are all 20% cement:

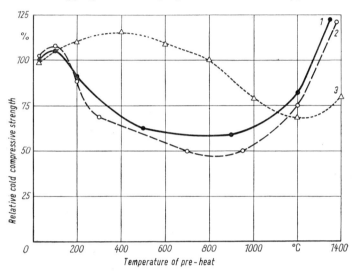

Fig. 45. Variation of the cold compressive strength of concrete made with barium aluminous cement with the temperature of pre-heat

1 Bac, chamotte; *2* Bac, magnesia; Bac = Barium aluminous cement; *3* the curve for high alumina cement-concrete is shown for the purpose of comparison (Refractory Industry German Democratic Republic)

80% aggregate. The obvious loss of strength to the extent of about 50% in the middle temperature range should be particularly noted.

As has already been seen in the case of other refractory concretes, the cold compressive strength after heating is dependent upon the quantity of cement in this case also. The strength of the concrete in the fired state increases in a similar manner to that of the unfired concrete.

5.1.2.2. Hot compressive strength

There is so far less experimental evidence available about the values of hot compressive strengths than about cold compressive strengths. This follows from the fact that certain experimental difficulties arise in the establishment of the former.

Measurements of strength by *W. C. Hansen* and *A. F. Livovich* [163] and by *H. Mitusch* [12] show that the hot compressive strength at first decreases, may then pass through an intermediate maximum in the region of 600 to 800 °C in a similar manner to the cold strength, and then falls still further. A few typical curves are repeated in Fig. 46. A further large decrease in strength must be expected beyond the temperatures to which the tests relate, since the values of binding strength are known to decrease with increasing temperature. For the rest, a comparison of the values in this diagram with those of the cold compressive strength given in Fig. 43 show that there are no large differences between the two, at least in the middle temperature range; the cold compressive

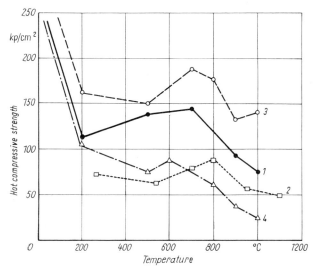

Fig. 46. Hot compressive strength of aluminous cement concretes (Mitusch, Hansen und Livovich)
1 Chamotte; *2* Chamotte; *3* Sillimanite; *4* Magnesia

strengths may be said to be about one-third higher than the values from the hot tests.[1] It can be concluded from these results that refractory concretes made with aluminous cements show satisfactory strength behaviour at high temperatures, provided that the proportions of cement and aggregate are correctly chosen and the right materials used.

5.1.2.3. Expansion and shrinkage behaviour

Conventional refractory materials expand with temperature in a reversible manner. The dimensions of expansion joints in structures made of such materials may be chosen, having regard to the known coefficients of thermal expansion, so that they close up when the plant is operating at its working temperature.

In refractory concretes the expansion processes are rather more complicated, because during the first raising to temperature the expansion of the refractory aggregate is superimposed upon a shrinkage of the binding agent constituent. The causes of this shrinkage in pure hydrated cements have already been dis-

---

[1] A number of investigations have shown [163], [290] that in parts of the middle temperature range the hot compressive strength may even exceed the strength in the cold condition.

cussed in detail in Sections 2.1.2 and 2.1.4 (see also Figs. 4 and 11). It will be evident that the total shrinkage of approximately 2% referred to there will not appear in concretes, since cement contents are of the order of 15 to 25% of the total mass of the concrete. The shrinkage effects are therefore decisively reduced and may even disappear altogether. Refractory concretes, like pure cements, expand like other materials, once they have been subjected to their first heating.

Although the shrinkage may thus be considerably reduced due to the small proportion of cement in the whole material, nevertheless the difference between the volumetric changes of the two principal constituents of the concrete would lead to the occurrence of considerable internal stress, if only the binding agent and ordinary aggregate were present. This would have a harmful effect upon the strength of the material. The very fine fraction of the aggregate has a very favourable influence in this respect, by appreciably reducing the shrinkage of the matrix containing the binding agent. Research by *K. D. Nekrasov* has shown that the effect of an aggregate such as chamotte dust, for example, can vary considerably, depending upon the type of the hydrated clinker mineral that is present. Alit shrinks some tenths of one per cent, belit only very little, while C_3A and C_4AF do not shrink at all, provided that the quantity of chamotte added is at least three times the quantity of cement. The last two, in fact, expand when first heated.

The effects of ceramic stabilisers in this regard are, however, by no means uniform. Research by *G. M. Ruščuk* [32] has shown that slags, pumice stone and fired clay certainly lead to reductions in the shrinkage of portland cement, but that trass, kieselguhr and siliceous materials, for example, increase the shrinkage, sometimes to an extraordinary extent (with kieselguhr and siliceous material). The effects of materials containing tuff are also extremely varied as a number of experimental results have shown. These marked shrinkage phenomena may be attributed to the high inherent shrinkage which the aggregates themselves exhibit when subjected to heat. In every case, therefore, tests must be made to ascertain how the aggregate behaves when subjected to increasing temperature.

In aggregates which have a positive influence, the reduction of shrinkage which is achieved is in general proportional to the amount of aggregate present. If the correct quantity of aggregate is used, shrinkage can be completely removed. The magnitude of the shrinkage will of course depend upon the type of aggregate as well as its quantity. Chamotte usually proves to be more effective than chrome ore, while brick dust is better than chamotte. The addition of loess has also been found to be extremely effective [164]. Quartz dust, if present in sufficient quantity, leads to a high rate of expansion under heat and also shows a sudden additional expansion at about 575 °C due to the β-α-quartz transformation. Thus the expansion and shrinkage behaviour of concretes can definitely be controlled and influenced by the fine fraction of the aggregates.

A number of typical expansion-shrinkage curves are given in Fig. 47. These have been obtained from various sources and relate to different types of fire-resistant concretes made with fine and coarse chamotte aggregates. The whole question of expansion and shrinkage behaviour is very complex and is affected by many factors, including not only the types of cement and aggregate but also the proportions of the various constituents, the grading and the particle size of the aggregates. The expansion which takes place upon heating will depend upon the concrete mix, but will sometimes be smaller than that of the pure fire-resistant material itself. An appreciable shrinkage component is often superimposed upon the expansion, which is thus rendered irreversible, so that the volume of the material after it has cooled down again is less than the

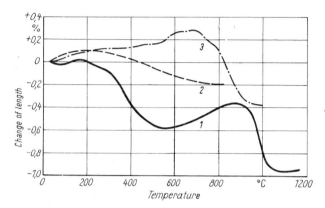

Fig. 47. Expansion/shrinkage behaviour of chamotte concretes made with various cements
(Nekrasov, Mitusch, Röhrs and Gibbels)

1 Aluminous cement + chamotte; *2* Portland cement + chamotte powder;
3 Blast furnace cement + chamotte

initial volume. These are, however, only transitory phenomena and are not of great importance if correctly proportioned concrete mixes are used, because after the first firing the concrete will have normal expansion characteristics. The problem becomes difficult only if appreciable shrinkage occurs during the first temperature excursion; cracks then appear in the refractory lining, which in normal fire-resistant materials will usually close up. The use of the right mix will, however, counteract this effect.[1]

The shrinkage of heat-resistant concretes made from portland cement between temperatures of 800 and 1000 °C is generally taken to be linear, with a value of from 0,2 to 0,7%. High quality portland cement concretes made with chromite sometimes exhibit a larger value of shrinkage at a somewhat higher temperature. The linear shrinkage of concretes made from aluminous cements is between 0,2 and 0,8%, depending upon the composition, and occurs up to 1100 °C. At higher temperatures there may even be expansion [166].

5.1.2.4. Porosity

In furnace linings the lowest possible porosity is usually desired on the hot face, for a number of reasons. In this connection, therefore, it is important to

---

[1] An interesting possibility of compensating for shrinkage exists in the use of small quantities of raw cyanite as an additive; when heated this forms mullite, resulting in volumetric growth [165].

know how far the transformations (or phase-changes) that occur during the
first firing are associated with a change (that is an increase) in the porosity.
Porosity changes of this type are only to be expected, since additional pores
are produced as a result of the expulsion of the water of hydration. An increase
in the porosity already occurs at the stage at which the strength falls off. Measurements of water absorption capacity, porosity and density all confirm this.

In normal refractory, dense (not lightweight) concretes, the density decreases
and the porosity increases as the temperature is increased. The porosity is

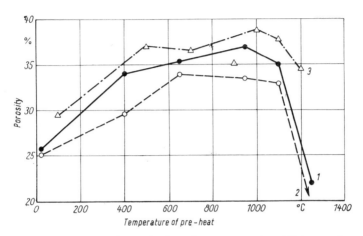

*Fig. 48. Variation of porosity in refractory concretes with temperature
of pre-heat (Gibbels, Schmid)*

1 Portland cement + chamotte; *2* Blast-furnace cement + chamotte;
3 Blast-furnace cement + chamotte + clay

a maximum in the region of minimum strength; *Nekrasov* gives figures of 20
to 30% (open porosity OP), with a water-absorption capacity (WA) varying
between 10 and 20%. It is self-evident that here also the values will vary
according to the method of manufacture, the type of the constituent materials
and the initial porosity of the unfired concrete. As an example, a maximum
open porosity of 21 to 24% is permitted in precast elements of refractory concrete made with cement and aggregates having an average content of Al_2O_3
[167]. The manner in which the porosity varies with the temperature of firing
can be clearly seen from the values given in Fig. 48, which are quoted from
L. Schmid [104] and *H. Gibbels* [55]. The increase in the porosity in the critical
temperature range is in some cases considerable, and therefore has an effect upon
the strength. The porosity will diminish to a greater or a lesser extent at high
temperatures depending upon the ease with which the materials can be sintered; the formation of smelted material can, however, in certain circumstances,
lead to the development of additional closed pores.

In practice, too much importance need not be attached to the maximum
value of the porosity, since it occurs only during the first heating of the concrete;
care should, however, be taken to ensure that the operating temperatures lie

above this maximum, and that the porosity decreases again. It is not likely that serious chemical attack, which is liable to be assisted by the presence of a large number of pores, will take place at relatively low temperatures in the range 800 to 900 °C. At higher temperatures such chemical action can, however, become a real problem, if the stage of maximum porosity has not been clearly passed. Here it is once again evident that the properties and the uses of refractory concretes must always be considered together. For medium operating temperatures a concrete that sinters at an early stage is thus preferable to one that sinters very completely, if no steps are taken to ensure that the latter is fired initially to the requisite extra high temperature.

5.1.2.5. Elasticity

The elastic behaviour of refractory concretes is of great importance, having regard to the many different types of structure in which they may be employed and the large range of possible operating temperatures. Of particular importance is the way in which this property varies with temperature. In the Soviet Union particularly, much fundamental research has been carried out in this field.

The modulus of elasticity of portland cement and aluminous cement concretes is generally quoted as being in the range 300000 to 400000 kp/cm² (4250000 to 5700000 psi). The modulus of elasticity of the concrete mass decreases sharply with increase of temperature. *Nekrasov* and his colleagues have established values of the decrease for portland cement concretes from 25 to 5% of the modulus in the unfired state, for concrete which is heated to 800 °C. The plastic component of the deformation increases at the expense of the decrease in the elastic component. This same phenomenon, or something very similar to it, holds for all other types of concrete. Up to about 800 °C, however, a predominantly elastic deformation is to be expected; above this temperature level it changes to become increasingly plastic in nature, the change being at first gradual, then increasingly more marked. Calculations that have been made to ascertain the plasticity index for numerous high temperature concretes all point towards this (the plasticity index is the ratio of plastic deformation to total deformation). Detailed descriptions of elastic/plastic effects in relation to the heating up of the concrete may also be obtained from a publication by *B. A. Al'tšuler, G. D. Salmanov* and *A. P. Tarasova* [168].

S. J. Schneider and *L. E. Mong* [169], amongst others, have carried out measurements of the modulus of elasticity and its relationship to temperature for concretes made with high alumina cement and chamotte and tested in the unfired condition. The results are shown in Fig. 49 and indicate the decrease of the elastic modulus that has already been described. The elastic modulus of the concrete, which has been fired at a temperature of 1050 °C, can be seen to remain fairly constant over the whole temperature range. Pre-heating to a temperature of 1300 °C raises the modulus again, presumably through the formation of new mineral phases and recrystallisation processes [170]. The same authors also found that the moduli of elasticity of concretes containing 20% cement were higher than those of concretes containing 10%. A cement content of 30% sometimes resulted in a further increase, sometimes in a reduction of the modulus, presumably because the reactions with the aggregates were more marked, or because on occasions there was an increase in porosity, resulting in a decrease in the elastic modulus.

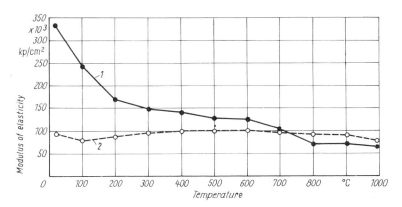

Fig. 49. Variation with temperature of the modulus of elasticity of aluminous cement concrete (Schneider and Mong)

1 unfired; *2* fired at 1050 °C

5.1.3. Thermal-mechanical Properties

This class of properties includes all those which are defined in terms of temperature or of mechanical properties which are temperature-dependent. The following relates principally to the softening and general strength behaviour of the concretes at high temperatures and under a great variety of conditions. The chief thermo-mechanical properties are defined as fire-resistance, softening under pressure, hot compressive strength, creep resistance, creep behaviour, and stability under fluctuating temperature. These are discussed below for each different type of refractory concrete.

5.1.3.1. Fire resistance

Refractory concretes are heterogeneous solid mixtures, whose constituent materials have very different separate fire-resistances. While the cone fusion test points of fire-resistant aggregates are generally above about 1630 °C, as already seen, those of the normal binding agents lie generally in the range 1250 to 1500 °C, depending upon the composition. The values for special cements, however, may be as high as about 1800 °C. The binding agent is therefore in most cases the most easily smelted constituent and acts as a flux at first, until new fire-resistant mineral phases have formed as a result of the chemical transformations described above. It is also worthy of note that portland cement, although normally the more resistant to smelting, may, if combined with certain aggregates (such as chamotte, for example) become a stronger flux than the more easily smelted normal aluminous cement. This will be seen when the fire resistances of the relevant refractory concretes are considered.

It will be easily seen that the cone fusion test point temperatures depend very much upon the type of cement and of fine and coarse aggregates and more especially upon the cement/aggregate ratio. The more fire resistant the aggregate and the higher its proportion in the concrete, the higher will be the cone fusion test point of the concrete. It is obviously not possible to give figures that apply in all cases, but it can in general be assumed that blast furnace cement has the lowest fire resistance and that this property increases through portland cement and normal aluminous cement to high alumina cement and barium aluminate cement. Aggregates may be arranged in order of increasing temperature of the cone fusion test point as follows: non-fire-resistant aggregates, middle quality and high quality chamotte, sillimanite, corundum, chromite and magnesia. If refractory materials are used as aggregates, it is generally true to say that the fire resistance of the concrete decreases with increasing cement content [6], [11], [13].

A selection of typical values for the cone fusion test point is given for various heat-resistant and fire-resistant concretes in Table 11. These have been abstracted from the published literature. It is of course possible to provide only a very general summary here, but the table does give a general indication of the order of fire-resistance values for the different types of concrete.

Table 11. The fire-resistance (cone fusion test point) of cement-bound high-temperature concretes

Cement	Very fine aggregate	Aggregate	Mix proportions	Cone fusion test point [°C]	Source
Blast furnace cement	Clay	Chamotte	15:10:75	1350	[55]
Portland cement	Tripoli powder	Chamotte	12: 8:80	1440	[110]
Portland cement	Chamotte	Chamotte	18: 6:76	1520	[6]
Portland cement	Chamotte	Chrome ore	15: 5:80	> 1750	[6]
Aluminous cement	—	Chamotte	15:85	1580	[11]
Aluminous cement	—	Sillimanite	20:80	1550	[12]
Aluminous cement	—	Corundum	20:80	> 1650	[6], [12]
Aluminous cement	—	Chrome ore	15:85	1800	[6], [108]
Aluminous cement (white)	—	Chrome ore-magnesia	20:80	1710	[13]
High alumina cement	—	Chrome ore-magnesia	20:80	1800	[13]
Barium aluminous cement	—	Chrome ore-magnesia	20:80	1880	[13]

5.1.3.2 Stability under compressive stress at high temperatures

A property which is of more practical importance than the cone fusion test point is the index of softening under pressure (which may also be described as the "stability under compressive stress at high temperature"). The tendency to soften under compressive forces is one of the most important properties to be considered in a refractory material, since it will normally be subjected to compressive stress simultaneously with heat (for example, due to its own self-weight). A material operating under this combination of conditions will start to deform much earlier than would be indicated by its cone fusion test point. The stability under compressive stress at high temperature is thus a measure which indicates approximately the allowable operating temperature.[1] However, the pressure of 2 kp/cm^2 (28 psi) used in the test is in excess of the stresses that normally exist in a refractory lining. In other words the test conditions are considerably more severe than the operating conditions; this means that the test piece starts to soften under the effects of pressure at a temperature lower than that at which softening would commence in practice. One might try to set the operating temperature at a somewhat higher level, if it were not for the time factor. This is not taken account of in the test procedure, but is a decisive factor in practical refractory technology. (This is discussed in greater detail in Section 5.1.3.4.)

The practice in the German Democratic Republic and in most other countries is to test the deformation of standard test pieces at specified elevated temperatures under a compression of 2 kp/cm^2 (28 psi). Characteristic curves are obtained from these tests, which show a similar form for all refractory materials, though each one will have a different scale (see Fig. 50). There is unfortunately no universal procedure for interpreting the curves and so establishing fixed points for the definition of the softening behaviour. In the German Democratic Republic the definitions t_a and t_e have been adopted. t_a defines the temperature at which the loaded test piece is compressed by 0,6% from the point of maximum expansion; t_e is the temperature at which the test piece is deformed by 20%.

The Czechoslovakian codes define t_e as the temperature at which a 40% deformation takes place. Soviet practice is to note the commencement of deformation and also in particular, the temperature at which 4% deformation occurs, 40% ($= t_e$), and the temperature of destruction. Since the first named of these points do not completely correspond to t_a, it is difficult to make a comparison of the softening figures; the orders of magnitude of the relevant temperatures do not vary to a great extent.

In principle the same factors that are decisive in relation to fire-resistance are also the important ones when stability under compressive stress at high temperature is being considered. In the case of some types of concrete there are, however, certain differences. For a proper understanding of these it is necessary to consider them in relation to the composition of the concrete.

5.1.3.2.1. Heat-resistant Concretes made with Slag Cements

Heat-resistant concretes of this type have a relatively low stability under stress at high temperature. This is inherent in their composition and their use in practice is determined accordingly. Research by *H. Gibbels* [55] has shown that the t_a-values for concretes made with blast-furnace cement and chamotte (cement content 15 to 30%) lie between 1070 and 1140 °C, and the t_e-values

---

[1] This is true within limits, since it depends very much upon whether the material is thermally loaded on one side only or on both sides; in the first case the allowable operating temperatures are higher.

between 1130 and 1220 °C. The softening range is not very large, usually being about 60 to 80 °C (see also Table 12). Refractory concretes were tested by *I. Schmid* [104], made with iron portland cement and blast-furnace cement, having clay dust and chamotte flue-dust as the stabiliser and shale chamotte as the principal aggregate. These proved to have t_a-values of 1150 to 1200 °C, and t_e-values (for 10% deformation) of 1230 to 1260 °C. It should be noted in this connection that there is some improvement in regard to softening under pressure if the concrete is fired several times at high temperatures; for example in one case the t_a-value was increased from 1190 °C (after one firing) to 1240 °C after the concrete had been fired seven times at 1100 °C. There is a deterioration in the resistance to softening under pressure, if normal gravel is used instead of chamotte. *Schmid* established for this case t_a-values of about 900 °C and t_e-values of 1100 °C. However, after the first annealing process, t_a increases to about 1100 and t_e to about 1180 °C.

5.1.3.2.2. Refractory Concretes made with Portland Cements

The temperatures at which portland cement concretes soften under pressure vary between extraordinarily wide limits, depending upon the type of ceramic stabiliser and aggregate used. It is therefore almost impossible to state generally applicable figures. Concretes of this type offer a very good example of the way in which the thermal properties of a concrete can be influenced or varied and of how a refractory material of high quality can be made, although a binding agent of relatively low fire resistance is used, provided that the aggregates are correctly chosen.

Various values of t_a and t_e for portland cement refractory concretes, drawn from published literature, are given in Table 12. The following tendencies may be deduced from this table: the resistance of the concrete to softening under pressure is generally improved by the use of aggregates of high refractory qualities; this is not, however, the only influencing factor, since the grading and shape factor of the aggregate may also have an appreciable effect. The reactivity of the aggregate is also evidently of importance, on account of the phase building processes that depend upon it and their effect upon the resistance to compression at high temperatures. If these conditions are assisted both by the chemistry and by the grain size, then the result is usually an increase in the temperature at which softening under pressure starts to take place.

A series of investigations by *G. D. Salmanov* [149] using portland cements from various sources, has shown that the degree of lime saturation of the clinker has hardly any effect upon the softening of refractory concretes at high temperatures, but that, on the contrary, the absolute contents of alumina and of ferrous or ferric oxide reduce the stability under compressive stress at high temperatures. This will be evident from the discussion in Section 2.1.2.2 about the softening of cements. The cements known as white cements are therefore basically better than normal cements. The most recent experience does show, however, that this disadvantage of mass-produced cements can be compensated by the choice of suitable aggregates. It will be seen from the technical references quoted [171] that t_a-values greater than 1400 and up to 1500 °C can be achieved by these means, giving fire-resistant and highly refractory concretes.

Table 12. Temperatures of softening under pressure for cement-bound high-temperature concretes

Cement	Very fine aggregate	Aggregate	Mix proportions	t_a or 4% deformation temperature [°C]	t_e [°C]	Source
Blast-furnace cement	Clay	Chamotte A I	15:10:75	1140	1220	[55]
Blast-furnace cement	Clay	Chamotte	20: 8:72	1170	1250	[104]
Portland cement	Clay	Chamotte	20: 7:73	1200	1240	[103]
Portland cement	Clay	Chamotte	18: 8:74	1220	1270	[104]
Portland cement	Magnesia	Chrome ore + quartz	1:2:0,3:0,2	1400	1510	[149]
Portland cement (containing P_2O_5)	Chrome ore magnesia	Chrome ore	10:20:70	1500	1600	[171]
Aluminous cement	—	Chamotte	15:85	1330	1450	[11]
Aluminous cement	—	Chrome ore	30:70	1330	1500	[149]
Aluminous cement	—	Sillimanite	20:80	1350	1505	[12]
Aluminous cement	—	Corundum	20:80	1480	> 1650	[12]
Aluminous cement	Magnesia	Chromite	15:25:60	1350	1430	[108]
High alumina cement	—	Chrome ore	≈ 20:80	> 1480		[136]
High alumina cement	—	Corundum	≈ 20:80	> 1550		[136]
Barium aluminous cement	—	Forsterite	15:85	1360	1520	[68]
Barium aluminous cement	—	Mullite	15:85	1460	1650	[68]
Barium aluminous cement	—	Corundum	20:80	1800		[68]

Chrome ore is an example of a constituent material that was used for this purpose.

It must be concluded from the results of numerous experiments that softening under pressure, like fire-resistance, is dependent upon the cement content, to the extent that increases in the cement content usually lead to a deterioration in the ability to withstand compressive stress at high temperature. This is only true, however, if the aggregates used are refractory materials; if they are not, the relationship stated may be reversed within certain limits.

The quantity and the fineness of the ceramic stabiliser also have an influence. This has already been referred to above. The factors which have favourable effects are chemical as well as purely physical. On the chemical side a higher degree of fineness may be expected to produce more rapid and more complete chemical transformations. The physical aspects, in the view of *Salmanov*, relate principally to the capillary forces which operate in the smelted phase between the refractory particles; with a higher degree of fineness the surface effects of the smelt will hold the whole heterogeneous mixture together more firmly, resulting in an improvement in the stability under compressive strength at high temperatures.

Nekrasov [6] states that in concretes made with portland cement and without basic or neutral aggregates, the commencement of softening under pressure occurs at 1100 to 1200 °C. The t_a-values will lie between 1250 and 1350 °C. The temperatures in concretes containing chrome ore and/or sintered magnesia will be up to 200 degrees higher. In the TGL and other standards used in the refractory industry in the German Democratic Republic, a minimum t_a-value of 1030 °C is required for portland cement concretes with chamotte aggregates (for concrete having a chamotte of medium grain size) and 1050 °C (for concrete containing coarse chamotte) [5], or alternatively 1100 °C [167]. For lightweight refractory concrete [172] the guaranteed values are: t_a 1050 °C, t_e 1150 °C. These portland cement-chamotte refractory concretes therefore correspond approximately, in their stability under compressive strength at high temperatures, to the chamottes listed under Type IV in TGL 4323. If special refractory aggregates are used, such as sintered magnesia, chrome ore or chrome ore-magnesia, the softening temperatures of good quality chamotte may be achieved. It has been shown by *G. D. Salmanov* [149] and *V. F. Guljaeva* and *G. D. Salmanov* [171] that, in special circumstances, values as high as 1500 °C for t_a and 1600 °C for t_e may be attained. There is a fundamental difference between the concretes and the chamottes, however, in that the range over which softening takes place is relatively short in the concretes. While the difference between t_a and t_e is at least 200 degrees for all chamotte bricks, the difference for refractory concretes is at most 100 to 120 degrees, and is usually less (see also Table 12).

5.1.3.2.3. *Fire-resistant Concretes made with Aluminous Cements*

The temperature for softening under pressure for a number of aluminous cement concretes are given in Table 12. Some typical curves of the properties of a refractory concrete of average quality are given in Fig. 50. The characteristic curves for normal chamotte-, silica- and magnesia-bricks are also given for the purpose of comparison. It will be noted that a distinguishing feature of concretes of this type is that their t_a-value lies close to that of chamotte but that the temperature difference between t_a and t_e is still comparatively small, although rather larger than that of the portland cement concretes. In many of the concretes the difference is so small that their curves of softening lie close to that of silica.

It will be noted that in the case of aluminous cement concretes, the temperatures at which softening takes place vary over a wide range, depending upon the cement content and the type of fire-resistant aggregates used. It should be mentioned, in amplification of Table 12, that various Soviet authors (quoted by *Budnikov* [10] and *Nekrasov* [6]) state t_a-values (for 4% deformation) as follows: chamotte concretes, about 1280 °C; chromite concretes, 1350 °C; and dunite concretes 1300 to 1380 °C. The corresponding t_e-values are given as 1350, 1410, and 1380 to 1420 °C, respectively. *E. Kuntzsch* and *G. Rabe* [128]

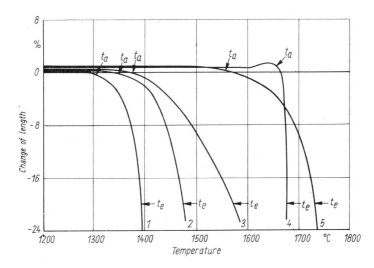

Fig. 50. Graph showing softening with temperature of refractory concretes made with aluminous cement (Zapp and Dramont) in comparison with conventional refractory bricks

1 Aluminous cement-chamotte concrete; *2* Aluminous cement-chamotte-corundum concrete; *3* Chamotte brick; *4* Silica brick; *5* Magnesia brick

report t_a temperatures of 1310 °C for smelted aluminous cement concretes with chamotte aggregates and t_a-values of 1420 °C and t_e-values of 1460 °C for concretes with sillimanite. The standards of the German Democratic Republic require minimum t_a-values of 1080 or 1280 °C, depending upon the density of the concrete [5], [167], [173]. For lightweight refractory concrete [172] a t_a-value greater than 1100 °C and a t_e-value greater than 1220 °C are specified.

The stability of the concrete under compressive stress at high temperatures deteriorates in general somewhat with increasing cement content. In many cases, however, the difference are not very important. Some exceptions to this rule have been shown by the investigations of *H. Mitusch* [12] to be corundum and certain types of sintered magnesia and chrome ore, for which the optimum hardness at high temperatures is given by certain values of the content of smelted aluminous cement. If non-refractory aggregates, such as broken brick, are used instead of refractory aggregates, then the temperature at which softening takes place increase with increasing cement content (*Salmanov, Mitusch*)[1].

The chemical nature of the aggregate appears to play an important part. It has been shown by *H. Mitusch* [12] that when aluminous aggregates are used, the resistance to compressive stress at high temperatures increases to a marked extent with increase in the content of Al_2O_3, and decreases with basic and neutral aggregates. He attributes this particularly to the variations in chemical reactivity as well as to the refractoriness of the aggregate. This view is supported by the findings of *Salmanov*, amongst others. This research worker found that an increase in the amount of chromite did not give the expected improvement in resistance to stress at high temperatures.

The grading of the aggregate also has quite an appreciable effect upon the tendency to soften under compressive stress. An increase in the proportion of fines usually improves the concrete in this respect, probably as a result of an increased readiness to react chemically. This is especially true of chrome ore, where the use of a very fine fraction and the achievement of more effective reactions through an increase in the period of reaction, can raise the t_a and t_e values to a noticeable extent.

If also the tests to ascertain the stability under compressive stress at high temperature are carried out on unfired test pieces, then it is important, in order to establish their validity, to know whether further chemical transformations will take place under the normal working temperatures, leading to possible modification in the softening properties of the concrete. A study of the technical literature has shown that experience with fire-resistant concretes made with normal industrial aluminous cements does not give uniform results. Soviet experiments with chamotte concretes [31], [60] have indicated that previous heat treatment of the test pieces did not have any appreciable effect upon the results; the t_a-values of fired samples were slightly higher than those of unfired samples. A large number of experiments by *F. Zapp* and *F. Dramont* [11] testified to a similar tendency in many cases, though sometimes the differences proved to be rather larger. There were, however, some test pieces which showed a drop in the t_a- and t_e-values after firing. The reasons for this have not yet been clarified. Similar observations have been reported by *H. Mitusch* [12]. For concretes with chromite aggregates a noticeable increase in the temperatures at which softening under compression takes place has been observed on fired samples, t_a being raised by at least 60 to 70 degrees. These experiments were conducted by *A. I. Rojzen* [31]. In some cases, however, the influence was smaller.

Some typical curves indicating resistance to compressive stress at high temperatures for fire-resistant concretes made with high alumina cement (SECAR 250) and aggregates of various types are shown in Fig. 51. These are quoted from *J. Arnould* [63].

Data relating to the resistance to stress at temperature of fire-resistant concretes made with barium aluminate cements have been published by *P. P. Budnikov* and *V. G. Savel'ev* [65] and *V. G. Savel'ev* [69]. Some figures are quoted in Table 12. The property is in this case very dependent upon the type

---

[1] The lowest values of *ta* in this system lie between 860 and 1040 °C. With a smelted aluminous cement content of 50% they rise to 1120 to 1190 °C. Proportions such as these are naturally of no practical use.

of aggregate used. When chamotte and basic aggregates, such as chrome-magnesia, dunite, and forsterite are employed, t_a-values of only 1230 to 1300 °C can be attained and t_e-values of only 1380 to 1530 °C, while mullite and corundum give figures of 1460 and 1800 °C respectively for t_a, and 1600 and more than 1800 °C for t_e.

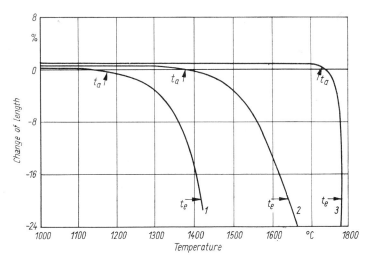

Fig. 51. *Temperatures of softening under pressure for refractory concretes made with SECAR high aluminous cement (Arnould)*

1 Chamotte concrete; *2* SiC-concrete; *3* Corundum concrete

5.1.3.3. Hot compressive strength

The term "hot compressive strength" is used to define the ability of a solid body to resist compressive stress at constant high temperature. It states the compressive stresses that are required, in order to destroy the material at a defined temperature.

The hot compressive strength of refractory concretes is rarely measured at temperatures as high as the highest experienced by the concrete in service. Its variation with temperature has already been discussed in detail in Section 5.1.2.2 so only a little remains to be added here.

The temperature-dependence of the hot compressive strength and the stability of the concrete under compressive stress at high temperatures are inter-related to the extent that the hot compressive strength must, by definition, have a value of approximately 2 kp/cm² (28,4 psi) at the t_e-temperature. The relationships become very confused, however, due to the overlapping of brittle fracture and plastic-viscous flow at high temperatures. The relationship between the temperature of softening and the strength at working temperature is normally related to the curve of hot compressive strength. Unfortunately, however, these relationships are not so well defined in the case of refractory concretes as they are for other fire-resistant materials; this is due to the intermediate phase transformations that take place in concrete. The tests to establish the hot compressive strengths of smelted aluminous concretes which have been made by *H. Mitusch* [12] extend only as far as 1000 °C; at this temperature the values obtained range from about 50 kp/cm² (710 psi) (for sintered magnesia aggregate) to about 150 kp/cm² (2130 psi) (for sillimanite aggregate). Using interpolation between the curves, in the 700 to 1000 °C range and the t_e-temperatures it is possible to estimate values of the hot compressive strengths for a temperature of 1250 °C. This method gives values of 30 to 60 kp/cm² (426 to 852 psi) for chamotte and sillimanite concretes respectively, having a cement content of 10 to 20%. The measurements of hot compressive strength obtained by *W. C. Hansen* and *A. F. Livovich* [163] gave a value of about 50 kp/cm² (710 psi) at approximately 1100 °C.

It has already been stated (see Section 5.1.2.2) that in the middle temperature range the hot and cold compressive strengths of refractory concretes correspond to a certain degree. The correspondence is so close in portland cement concretes containing ceramic stabilisers, that, as *K. D. Nekrasov* [174] has shown, there is a temporary increase of the hot compressive strength after the critical temperature range has been passed. If tests were carried out at still higher temperatures, the strength would presumably start to fall at a little above 1100 °C.

Values of the hot compressive strength at 1000 °C for the refractory concretes manufactured by the VEB Chamotte Works at Brandis are stated by *G. Franke* and *F. Kanthak* [107] to be 60 to 80 kp/cm² (852 to 1136 psi) for portland cement and 80 to 100 kp/cm² (1136 to 1422 psi) for aluminous cement.

5.1.3.4. Creep resistance

For the practical applications to which fire-resistant materials are put, creep resistance is of more importance than fire resistance, stability under compressive strength at high temperatures, or hot compressive strength. A refractory concrete is not usually taken to the limits of its thermal resistance, nor is it subjected to increasing compression at any fixed temperature until it fails in compression; it is, however, subjected to constant compression at its operating temperature extending over a long period of time. This time factor is not taken into account at all in the majority of materials tests, although its effects are of the greatest importance; in many cases, therefore, the results of tests give only very general indications and tell us nothing about the actual thermomechanical behaviour of the material under operating conditions. For this reason the long-term behaviour is now becoming a subject of experimental investigation to an increasing extent — at least in scientifically oriented circles. Long-term testing is desirable in those materials which, like refractory concretes, do not develop certain properties until after they have been fired and which consequently continue to have varying characteristics, as a result of internal reactions, until a final steady state is reached.

There is no uniform, established system for carrying out such tests. In association with the test for resistance to compressive stress at high temperatures, the variation in length is measured at a particular constant compressive stress at a particular constant temperature for a particular period of time. The creep behaviour of solid bodies — that is to say, the variation of shape under long-term loading — is thus ascertained from this test. All refractory materials have this property of creep; in the case of chamotte, for instance, it is already noticeable at a temperature of 900 °C.

The results of decisive experiments to establish values of creep resistance for fire-resistant concretes are very seldom published. Some figures have been given for concretes made with portland cement and aluminous cement, using chamotte and chromite aggregates respectively. These have been provided by *B. A. Al'tšuler, G. D. Salmanov* and *A. P. Tarasova* [168]. The tests lasted only one hour, however, and the maximum temperature involved was 1000 °C. The deformations are shown in Fig. 52. Fig. 53 gives the results of creep experiments

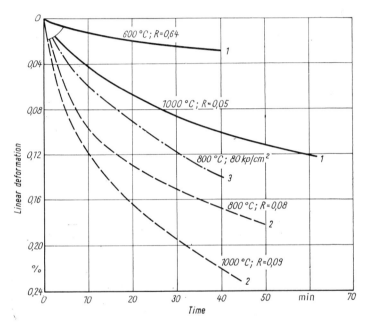

Fig. 52. Relationship between the deformation of portland cement and aluminous cement concretes and the temperature, loading and time (Al'tšuler, Salmanov and Tarasova)

The magnitude of loading *R* is the ratio of the pressure exerted to the maximum compressive stress at temperature *T*. *1* Aluminous cement + chrome ore; *2* Portland cement + chrome ore; *3* Portland cement + chamotte

on concretes made with high alumina cement (SECAR 250 and corundum) published by *J. Arnould* [63]. These illustrate clearly the practical usefulness of this high quality material.

Some conclusions about the creep behaviour can, however, be drawn from the other material properties. Thus it has already been seen, for example, that the tendency to soften under pressure is often reduced after firing. This means that in such cases the material must become more heat-resistant in a practical

10*

sense during the thermal treatment and therefore that the creep resistance is increased. The creep that was originally to be expected in the material should consequently take place at a slower rate, or even be eliminated altogether, at any rate for a limited period. Similar conclusions can be drawn from curves published by *H. Mitusch* [12] which show the relationship between hot compressive strength and temperature. These show that in many concrete mixes the strengths at 1000 °C are higher than at 900 °C.

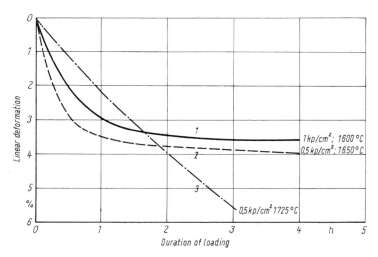

Fig. 53. Creep resistance of concrete made with high alumina cement and corundum (Arnould)

1 Normal corundum; *2* Normal corundum; *3* Special fused alumina

The large number of investigations that have been carried out by Soviet research workers into the elasto-plastic behaviour of concretes at high temperatures, which have been well described in Soviet technical literature [6], [168], may be compared in certain respects to creep tests, in that they include the study of plastic deformation at specified temperatures and under long-term loading of various intensities. Fig. 54 relates to the plastic deformation of portland cement/chamotte concrete and of smelted aluminous cement/chrome ore concrete, and depicts the relationship between stress and deformation for T (temperature) = constant, instead of indicating the more usual relationship between deformation and time with T and p (stress) both constant. It can be seen from this diagram that the plastic deformation, and therefore the compression, of the materials increases steeply with temperature and applied load. This means that that component of the plastic deformation which is responsible for the long-term deformation (or creep) does not really begin to be appreciable until a temperature of 1000 °C is reached, and does not even then reach a high value until a relatively high loading is applied. Permanent deformations do appear at temperatures in the middle range, but their magnitudes are negligible.

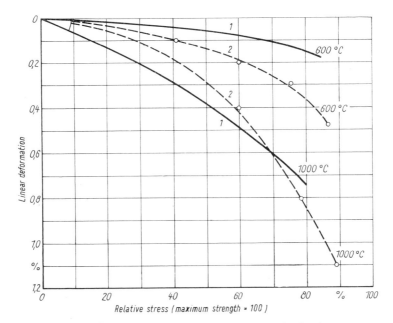

Fig. 54. Plastic deformation of portland cement and aluminous cement concretes in relation to temperature and loading (*Nekrasov, Al'tšuler, Salmanow and Tarasova*)

1 Aluminous cement + chrome ore; *2* Portland cement + chamotte

5.1.3.5. Stability under fluctuating temperature

The term "stability under fluctuating temperature" is used to define the ability of a solid body to withstand rapid variations of temperature without damage or destruction. The conditions of test and the methods used to assess this property are more varied and less standardised than those of almost any other type of test. Comparisons between different published results may therefore be made only with certain reservations. Full information about the test procedure and conditions is essential in every case.

Stability under fluctuating temperature is a typical example of a material property which has so far been determined by conventional methods; these methods do not have any physical basis. The usual practice is to heat testpieces, either completely or on one side only, to a specified temperature (850 or 950 °C) and then to chill or quench them in a cold air-blast or in water. The number of quenchings sustained before either the first damage occurs, or before a specified weight of the test piece spalls off, is used as a measure of the material's stability under fluctuating temperature.

This property has been assessed from completely new standpoints by *Th. Haase* [175]. Using the relationship between the maximum strain capacity of the material ε, $(= \sigma_z/E)$ and its coefficient of thermal expansion, he deduces the following expression for the temperature difference which can be sustained without damage:

$$\varDelta T = \frac{\varepsilon_{max}}{\alpha}.$$

Using this expression *G. Rudolph* [176] has studied the behaviour of refractory concretes made with smelted aluminous cement under fluctuations of temperature; this is discussed in more detail below.

Refractory concretes possess remarkably good stability under fluctuating temperature in comparison with all other refractory materials. This fact has been confirmed by all those who have studied the subject. Both in laboratory tests and in practical applications refractory concretes can normally withstand a large number of thermal shocks. They excel over normal refractory materials in this respect.

K. D. Nekrasov [6] reports that concretes made with portland cement and chamotte will sustain from 10 to 15 quenchings before the first hair cracks appear, and from 20 to 25 quenchings before the occurrence of the first fissures (or real cracks). At this stage the cold compressive strength still has an average value of 75 to 80% of the initial compressive strength. Concretes which contain slag or chrome ore cannot support so many repetitions of quenching.

A. Braniski [13] found that portland cement/chamotte concrete and bauxite concrete could support between 47 and 64 quenchings before destruction occurred. For refractory concretes made with blast furnace cement, *H. Gibbels* [55] has found that the number of quenchings generally exceeds eight.

H. Mitusch [12] and *A. Braniski* [13] have made a large number of systematic tests, using the quenching method, to establish the stability under fluctuating temperature of aluminous cement refractory concretes. The numbers of quenchings for concretes with aggregates containing 80% of chamotte, sillimanite or bauxite were respectively in the ranges 45 to 85, 70 to 90, and 45 to 50. The numbers were considerably smaller for aggregates consisting of corundum, sintered magnesia, chrome ore and chrome ore magnesia; the values in these cases reached only 8 to 30, 2 to 3, 6 to 10, and 9 to 11, respectively. It is reported by *Braniski* that chamotte concretes made with special cements of high alumina content reached a figure of more than 80 quenchings, while the figure for similar concretes containing chrome ore magnesia aggregates was only about 10. Similar results were obtained with the use of barium aluminate cement; with chamotte aggregate the figure was almost 90, but with chrome ore magnesia aggregate only 13.

As in the case of all other concrete properties, the stability under fluctuating temperature is a function of a large number of factors. Among the most important are the coefficient of expansion of the materials, the aggregate grading and the porosity of the concrete. The strengths of bonding and the temperatures of firing also play an essential part. Systematic observations in this regard have so far, however, rarely been undertaken. The investigations usually relate principally to the type and quantity of the aggregates and to the influence of pre-treatment and firing.

In so far as the aggregates are concerned, it is always beneficial to the finished concrete to use an aggregate which itself tends to be stable under fluctuating temperature, especially if this property results from the expansion characteristics of the aggregate. The contrast between chamotte and sillimanite on the one hand and corundum and magnesia on the other exemplify this. The facts quoted above show that the differences are great. High chemical reactivity of the constituent materials, leading to firm ceramic bonding, will obviously be conducive to resistance to thermal shock. The higher quality of concretes made with aggregates containing alumina, in contrast to those made with magnesia, chrome ore magnesia, and chrome ore indicate this. A high proportion of fines in the aggregate appears to be, generally speaking, unfavourable, because presumably a dense material structure is the result. This is known to be a factor leading to sensitivity to temperature changes.

The observations with regard to the effect of the proportion of cement in the mix are not consistent. While *A. Braniski* [13] found that there was a clear and reproducible increase in the number of possible quenchings when the cement content was increased from 10, through 15 to 20% in barium aluminate concretes, *H. Mitusch* [12] was unable to demonstrate any similar unambiguous tendency in concretes made with normal smelted aluminous cement. In the case of corundum and sintered magnesia aggregates the number of quenchings increased with the cement content, in the case of chamotte and sillimanite it decreased, while with chrome ore it remained constant.

The stability under fluctuating temperature of portland cement/chamotte concretes has been shown by *A. E. Fedorov* [158] to be improved by steam treatment and autoclave treatment, a fact which can be attributed to the raising of the structural strength. In this case hardening under pressure at 170 °C for a period of 12 hours gave better results than simple steam treatment at 85 °C for 12 hours.

Pre-firing of the concrete also has a favourable effect upon the stability under fluctuating temperature, as a general rule [12]. The temperature must, however, be high enough to ensure good ceramic bonding. *G. Rudolph* has shown [176] that the stability under fluctuating temperature varies in parallel with the cold compressive strength. Different results are therefore obtained if a refractory concrete is quenched from different temperatures, since, as seen above (see also Fig. 44), the strength of a smelted aluminous cement concrete will usually reach a maximum at about 800 °C and a minimum at about 1000 °C. In these circumstances the results of tests undertaken in accordance with the Soviet procedure (quenching from 800 °C) would be expected to prove better than those obtained from tests conducted under the German system (quenching from 950 °C). In practice, therefore, concretes which have been raised to the upper limit of operating temperature should prove to be relatively more resistant when cooled, than those with an operating temperature somewhere around the minimum strength point in the lower temperature range.

The addition to the cement of substances containing boron was found by *O. K. Alešina* [151] to have a most beneficial effect upon the stability under fluctuating temperature of concretes made with portland or aluminous cements and chamotte. These concretes could withstand far more quenchings from 800 °C than similar boron-free concretes. This phenomenon is attributed to the increase of strength, which was observed both in the unfired state (contrary to previous opinions) and at 800 °C. The combination of the boron influences the mechanism and the chemistry of hardening, in that a smaller quantity of

free hydrate of lime arises; in addition the secondary chemical combination of
the lime appears to take place at much lower temperatures than normally,
leading to an improvement in the strength values and in the stability under
fluctuating temperature. Another factor is that concretes which contain boron
have a somewhat lower coefficient of thermal expansion than boron-free concretes, the difference being about 10×10^{-7} per °C.

The investigations already referred to carried out by *G. Rudolph* showed that
it was possible, using the formula stated by *Th. Haase* [175] and a method of
measuring the maximum strain evolved by *Th. Haase* and *K. Petermann* [177],
to make predictions about the stability of refractory concretes under fluctuating
temperature. From the linear coefficient of expansion α and the experimentally
determined maximum strain ε_{max} temperature differences may be calculated,
which the material will withstand without damage. The measurements of
maximum strain, as a function of the temperature of firing, are given in Fig. 55
for test pieces of aluminous cement/portland cement/chamotte concrete, together with the corresponding values of the cold compressive strength and the
calculated temperatures as a measure of the stability under fluctuating temperature. The parallelism referred to above is quite evident from this diagram.

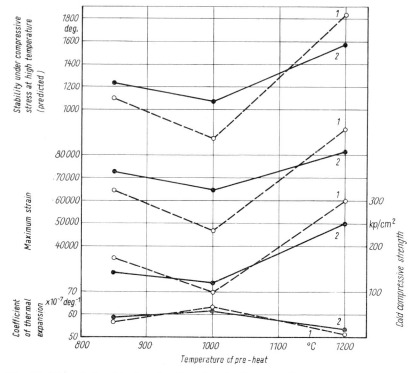

*Fig. 55. The relationship for aluminous/portland cement-chamotte concrete, between
the temperature of pre-heat and various thermal and mechanical properties
(Rudolph)*

1 20% Cement; *2* 25% Cement

The temperatures calculated are very high and lie well above the majority of those found by Haase and Petermann for other refractory materials. (Examples are: chamotte ≈ 300 to 600, silica ≈ 200 to 300, magnesia 16, chrome magnesia ≈ 500 °C). The outstandingly good stability under fluctuating temperature of refractory concretes which is demonstrated in practical operating conditions is thus confirmed by theory. The favourable strain values arise from the heterogeneity of the material and the distribution of the pores. The low coefficient of expansion in the case of chamotte also helps to improve the stability under fluctuating temperature.

A characteristic of refractory concretes is the mode of failure under fluctuating temperature. While normal refractory materials usually break up into lumps, refractory concrete slowly crumbles and splinters until it reaches the stage of complete disintegration. This phenomenon also can be explained in terms of the different structure of the concrete.

5.1.4. Physical and Mechanical Properties

A number of requirements can be set down and various values of properties may be specified for concretes which are to be used at high temperatures, which come under the heading of physical and mechanical properties. Among the most important are the density, porosity, thermal conductivity and thermal expansion. Some of these have already been discussed elsewhere, in connection with the irreversible changes which take place with temperature; that is during the transition from an hydraulically bonded to a ceramically bonded material. These properties will now be discussed in terms of values which are either independent of temperature (such as porosity) or where the temperature dependence is reversible (α, λ). In practice in the case of refractory concretes, therefore, they are properties which come into play either only in the hydraulically bonded state or only in the ceramically bonded state.

5.1.4.1. Specific gravity

The specific gravity of heat- and fire-resistant concretes are functions of their composition, the type of cement and aggregate, and the proportions of the mix. It is therefore self-evident that fixed values for specific gravity cannot be given. The lower limit for the specific gravity of a concrete is fixed by the specific gravity of the lightest constituent, in this case by the products of the reaction of the portland cement (for example, calcium hydroxide ≈ 2,25; okenite ≈ 2,3; gyrolite ≈ 2,4; tobermorite ≈ 2,45; afwillite ≈ 2,6; xenotlite ≈ 2,7) and by the aluminium silicate-containing aggregates, such as ash, chamotte, etc. with a specific gravity of about 2,65. The upper limit is determined by the heavy aggregates. Examples of these are: mullite and sillimanite ≈ 3,2; silicon carbide ≈ 3,2; corundum ≈ 4,0; magnesia (periclase) ≈ 3,6; magnoferrite and chrome ore ≈ 4,5.

Measurements by *H. Gibbels* [55] have established specific gravities for blast-furnace cement concretes of 2,45 to 2,6 in the unfired state and 2,7 to 2,8 in the fired state. The specific gravity of the concrete made by the Brandis works from smelted aluminous cement is stated by *G. Franke* and *F. Kanthak* [107] to be approximately 2,6.

5.1.4.2. Density

The density of refractory concretes is a function of the specific gravities and the densities of the constituent materials and of the volume of pores present in the concrete. The normal refractory concretes, other than lightweight concretes, are first considered here and values are stated for some of their properties.

It is a fundamental fact that figures for density will be appreciably lower than the corresponding values for specific gravity. The porosity of the concrete and also, in the case of pre-fired materials, the incomplete sintering are the principal reasons for this. Soviet sources [6] quote densities of 1,8 to 2,0 (112 to 125 lb/ft³) for chamotte concretes and 2,8 to 2,9 (175 to 181 lb/ft³) for chrome ore concretes. Measurements by *H. Gibbels* [55] have established values of about 1,7 to 1,8 (106 to 112 lb/ft³) for the densities of refractory concretes made with blast-furnace cement in the unfired condition and 1,65 to 1,75 (103 to 109 lb/ft³) for the same material after heating to 1000 to 1200 °C. Other authors state figures of 1,7 to 2,0 (106 to 125 lb/ft³) for portland cement concretes made with chamotte (fired), 1,6 to 1,9 (100 to 119 lb/ft³) for smelted aluminous cement concretes under similar circumstances and up to 2,0 (125 lb/ft³) when corundum is added [11]. For unfired smelted aluminous cement concretes with chamotte the values are about 1,75 (109 lb/ft³) [107], but occasionally also 2,1 (131 lb/ft³), while for concretes with sillimanite 2,85 (178 lb/ft³) is quoted [128]. The standards for the refractory concrete works in the German Democratic Republic [5], [167] require densities for chamotte concretes of normal composition of at least 1,65 to 2,05 (103 to 128 lb/ft³) when aluminous cements are used and 1,65 to 1,9 (103 to 119 lb/ft³) for portland cement. For precast components of average density the required values are 1,4 to 1,6 (87 to 100 lb/ft³) [173].

5.1.4.3. Water absorption, porosity and permeability

Density and porosity are related, being inversely proportional to each other. In practical applications, however, it is often not the total porosity of refractory materials that is important, but the ratio of closed pores to the open pores which are in contact with the atmosphere. A distinction has therefore been drawn between gross porosity (GP: hitherto usually described as true porosity) and open porosity (OP: previously defined as apparent porosity) (see for example TGL 9258 [178]). Since one of the methods of determining the open porosity is by measuring the water absorption, the water absorption (WA) is also usually stated.

The possible variations in porosity with temperature of firing have already been discussed in Section 5.1.2.4. The following is concerned principally with the initial values and with the values of porosity which exist after heating and, consequently, at operating temperatures.

The standards used in the Rietschen works [167] permit open porosities in unfired precast refractory concrete components of 24% for portland cement/chamotte concretes and 21% for smelted aluminous cement/chamotte concretes. Other factories produce mixes which sometimes have higher figures of porosity; for example, the Brandis works with 29% open porosity and 33%

gross porosity [107]. For portland cement and blast-furnace cement concretes *Schmid* [104] quotes average porosity figures (apparently gross porosity) of about 25%, while *Ludera* [103] reports 15 to 30% open porosity for portland cement concretes. *Gibbels* [55] found gross porosities of about 30% in blast-furnace cement concretes. The concretes which were investigated by *Zapp* and *Dramont* [11] are stated to have porosities of from 13 to 30%. Open porosities of between 18 and 25% are reported by *Braniski* [13] for his concretes, which covered a great variety of binding agents and aggregates. The figures for American refractory concretes lie in the range 20 to 30% [162], [179]. With regard to the relationship between gross porosity and open porosity it appears that in the case of refractory concretes, as with other refractory materials, the principle holds that the material is essentially open-pored.

The magnitude of the porosity after firing is a function of many factors, of which the most important are the temperature level and the reactions that occur between the binding agent and the aggregates. The information available on the subject therefore exhibits extraordinary diversity, containing examples of both reductions and increases in porosity. Thus for blast-furnace cement concrete *Schmid* [104] quotes a decrease from 25 to 11% after heating to 1200 °C while *Gibbels* [55] reports an increase from 30 to between 35 and 38%. Results from practical experience, however, show in most cases a decrease in porosity after treatment at high temperature [6; p. 141].

When assessing variations in porosity, however, regard should always be paid to the relationship between the temperature under consideration and the sintering temperature range of the concrete.

The porosity of fire-resistant dense concretes in the fired state is also a function of the water content among other things [179]. Thus it has been shown for American aluminous cement concretes that the open porosity after firing at 950 °C increased from 27 to 31%, if the water content was increased from 7 to 16%. The conclusion to be drawn from this is that the minimum water possible should be used in the manufacture of refractory concretes; this will, however, lead to difficulties in the handling and working of the concrete.

It is usually assumed that the open porosity of refractory concretes is about 10% higher than that of other refractory materials [103].

The values usually given for water absorption are in the range 15 to 20% [6], [10], [107], [136], but figures between 2 and 6% have also been reported [128]. In the case of cement mortars, without coarse aggregates, water absorption values in the region of 6 to 8% have been measured, but these were for material without any visible cracking.

With regard to the ease of passage of gases and liquids under the influence of pressure, the permeability of the concrete gives a better guide than porosity figures, although the latter, calculated from the water absorption or from the specific gravity and density, are commonly used. The permeability is chiefly a function of the diameter and length of the pores and, to a lesser extent, of the gross porosity. In only a minority of cases is there any relationship to porosity. Permeability is measured in the unit known as the Nanoperm (nP), though in technical literature the Darcy is also sometimes used (1 Darcy = 9,87 nP).

Normal refractory materials vary considerably in their permeability. Thus the values for chamotte lie between 0 and 400 centidarcies, those for sillimanite between 0 and 150, magnesite between 50 and 300, chrome magnesite between 250 and 1000 [1]. Tests by *G. R. Eusner* and *D. H. Hubble* [105] and by *G. R. Eusner* and *J. T. Shapland* [179] on refractory concretes made with aluminous cements have resulted in figures of from 0 to 80 centidarcies. The permeability to gases is only lower than that of normal refractory materials, however, if the concrete is not fired at temperatures in excess of 800 °C. Above this temperature the permeability increases sharply (see Fig. 56) and reaches the level

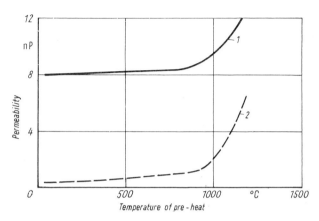

Fig. 56. Variation of permeability with method of placing and with temperature of pre-heat, for aluminous cement concretes (Eusner and colleagues)

1 trowelled concrete; *2* punned concrete

of permeability of normal refractory materials. In regard to permeability to gases, refractory concretes are therefore particularly suitable in the lower temperature ranges. The water/cement ratio has a very large effect upon the permeability, however; it has been shown that an increase in the water content from 7 to 16% can lead to a decrease in the permeability of an unfired commercial concrete.

It is possible, by means of treatment in an autoclave, to reduce the permeability to gases of a portland cement/chamotte concrete in the temperature range 20 to 400 °C effectively to zero, as has been demonstrated by *A. E. Fedorov* [158]. Normally hardened or steam-cured concretes still have a certain permeability under these conditions. These differences decrease considerably after firing at 800 °C.

5.1.4.4. Thermal expansion

Some aspects of thermal expansion have already been dealt with in connection with the irreversible changes of concrete properties that occur upon heating. After the high temperature reactions have taken place the irreversible expansion and shrinkage behaviour of refractory concretes is replaced by a normal reversible thermal expansion, which is in essence repeatable.

Since the cement content of a concrete is usually small in proportion to the aggregates, the coefficient of thermal expansion of concretes is largely determined by the expansion coefficients of their aggregates. Thus the coefficients of expansion of portland cement concretes made with various aggregates lie approximately in the following ranges:

Chamotte concrete	50 to $80 \times 10^{-7}/°C$	[6], [17]
Slag concrete	40 to $100 \times 10^{-7}/°C$	[17]
Chrome ore concrete	$60 \times 10^{-7}/°C$	[171]

They are less than those of the pure binding agent, which are given as about 100 to $130 \times 10^{-7}/°C$.

Similar values of the coefficient of expansion are valid for normal commercial aluminous cement concrete; for chamotte aggregate figures of $50 \times 10^{-7}/°C$ [107] and 50 to $65 \times 10^{-7}/°C$ [6], [176] are given. *J. F. Wygant* and *W. L. Bulkley* [160] established coefficients of around 40×10^{-7}, which are lower than those for chamotte, the difference being attributed to the heterogeneous nature of the concrete. Similar lower figures (on occasions much lower ones) are also reported by *F. Zapp* and *F. Dramont* [11].

The investigations conducted by *G. Rudolph* [176] show that the coefficient of expansion of concrete made with smelted aluminous cement is also dependent upon the temperature to which the concrete has been heated; all concretes made from chamotte showed a maximum value at about 1000 °C (cf. Fig. 55).

For corundum concretes made with cement of a high alumina content (SECAR; 72% Al_2O_3) coefficients of expansion of from 25 to $50 \times 10^{-7}/°C$ between 0 and 1400 °C are reported [63].

5.1.4.5. Specific heat

A knowledge of the specific heat of materials used in furnace construction is of importance when making thermal calculations. In the case of fire-resistant concretes the thermal capacity of the aggregates is usually the determining factor, since they constitute the principal component The specific heats of the respective aggregates may therefore be substituted without any appreciable error resulting; for example, chamotte ($\approx 0,20$ to $0,25$), magnesia ($\approx 0,10$ to $0,30$), chromite ($\approx 0,18$ to $0,22$), and so on (all values in cal/g °C). In general a figure of about 0,20 cal/g °C may be used for all refractory materials, other than carbon, at low temperatures; as the temperature increases the specific heat increases to about 0,28 to 0,30 cal/g °C at 1400°C

For exact calculations the characteristics of the separate aggregate materials and the effects of the binding agent and the products of chemical reactions naturally cannot be completely ignored; similarly attention must be paid to the formation of certain partly smelted compounds, since in this case the thermal capacity increases steeply.

Values of specific heat obtained from tests on different refractory concretes have been reported by *E. Hammond* [165]. Some of these are given in Table 13. The VEB Chamotte Works of Brandis give figures of 0,22 to 0,24 cal/g °C for their refractory concretes made with smelted aluminous cement.

Table 13. The specific heats of various high-temperature concretes (Hammond). Ratio of aluminous cement : aggregate ≈ 1:3 to 1:4

Aggregate	Specific heat [cal/cm³ deg °C]
Vermiculite	0,15
Diatomite	0,25
Expanded clay	0,35
Sillimanite	0,6
Chrome ore/Magnesia	0,6
Silicon carbide	0,7
Fused corundum	0,9
Fused magnesia	0,9

5.1.4.6. Thermal conductivity

The thermal conductivity of refractory materials is a complex property, depending upon the thermal conductivity of the separate constituents and also upon the grading of the aggregate, the volume of pores and the temperature. Viewed first from the point of the porosity, which has an important effect, especially in the case of lightweight concretes, it is found that the coefficients of thermal conductivity of refractory concretes are generally in the same region as those of their aggregates. Information about the values of the latter has already been given in Section 2.3.

The thermal conductivities of portland cement concretes in the temperature range 20 to 390 °C have been the subject of particular study and experiment by *Nekrasov* and his co-workers [6]. For the chamotte concretes used by the

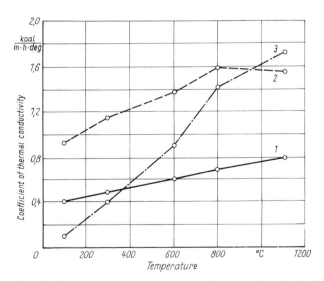

Fig. 57. Variation with temperature of the thermal conductivity of refractory concretes made with portland cement (Ludera)

1 Portland cement + chamotte;   *2* Portland cement + chamotte + clay; *3* Portland cement + clay

authors the values in the dried state lie in the range 0,35 to 0,62 and, after firing at 800 °C, in the range between 0,3 and 0,55 kilocal/m hr °C. For chrome ore concrete in the unfired state figures of 0,55 to 0,75 kilocal/m hr °C were established. Figures for thermal conductivity in the temperature range 100 to 1100 °C have been determined by *L. Ludera* [103]. The variation of this property with temperature for three different types of concrete is given in Fig. 57. Here it can be seen that the coefficient of thermal conductivity of these materials increases with increasing temperature; this is also true for a large number of other concretes.

For aluminous cement/chamotte concrete various figures are reported; e.g. 0,80 to 0,95 kilocal/m hr °C [12]; 0,6 [107]; 0,55 to 0,65 [136]; 0,80 [63]. Values of thermal conductivity for some concretes made from different types of aggregates are quoted from *E. Hammond* [165] in Table 14.

Table 14. The thermal conductivities of various high-temperature concretes (Hammond). Ratio of aluminous cement: aggregate ≈ 1:3 to 1:4	Aggregate	Coefficient of thermal conductivity [kcal/m · h · °C]
	Chamotte	0,7
	Chrome ore magnesia	1,0
	Sillimanite	1,25
	Calcined bauxite	1,5
	Sintered magnesia	1,5
	Fused corundum	2,0
	Fused magnesia	3,0
	Silicon carbide	≈ 6,0

In most cases the thermal conductivity of a concrete lies below that of its pure aggregates. In certain cases it may even be much lower, without the concrete necessarily being classed as an insulating concrete. This fact is explained in the technical literature on the subject as due to the higher porosity of refractory concrete when compared with other refractory materials. This is not, however, the case for all types of concrete; there must, therefore be other influencing factors, such as the very heterogeneous nature of concrete or the heat transfer between the different phases.

5.1.4.7. Resistance to abrasion and resistance to erosion

In many industrial uses refractory materials are subjected to erosion and wear. A good resistance to these effects is important in ensuring long life and is therefore extremely valuable in such cases.

That which is generally true for refractory materials is also of course true for refractory concretes. In many cases there is a certain amount of wear, and it is therefore necessary to be aware of the behaviour of the concrete in this respect.

The earliest investigations in this direction appear to have been carried out by *J. F. Wygant* and *W. L. Bulkley* [160] and by *W. B. Paul* [180] with particular reference to the conditions existing in petroleum refining. At a later date it was shown by *W. H. Gitzen*, *L. D. Hart* and *G. MacZura* [166] that concretes made with cement of high alumina content possessed particularly good resistance to erosion and surpassed other concretes in this respect. This was confirmed by *C. R. Venable* [181] by systematic erosion and wear tests on samples which had been previously heated to about 550 °C. Comparisons between chamotte firebricks and various chamotte fire-resistant concretes, which were made by *A. F. Milovanov* and *V. S. Zyrjanov* [182] showed that

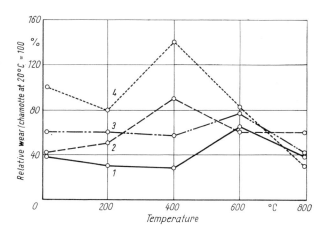

Fig. 58. Wear resistance of concretes made with portland cement, aluminous cement and waterglass, compared with chamotte bricks at various temperatures (Milovanov and Zyrjanov)

1 Waterglass concrete; *2* Aluminous cement concrete; *3* Portland cement concrete; *4* Chamotte brick

the wear of the concretes up to about 600 °C was less than that of the chamotte bricks. From 800 °C the resistance to wear was equal. The results which are of interest are shown graphically in Fig. 58. The concretes were tested at the relevant temperatures. The resistance to wear in the hot condition is noticeably better than after the concrete has cooled.

The relationships which American authors have found between resistance to abrasion and other properties are, unfortunately, ambiguous. A high density generally favours wear resistance, but there are exceptions. The same is true for compressive strength; good resistance to wear is usually, but not always, associated with high compressive strength. The resistance to abrasion increases with decreasing water content of the original mix, compressive strength increasing, of course, in the same sense.

It has been pointed out by *Wygant* and *Bulkley* that a distinction must be drawn between the effects of abrasion (or wear) and erosion. In the case of the first named it is the whole surface of the material that is attacked, while in the second case only those mineral phases which have the lowest strength are attacked. There is therefore a direct relationship between compressive strength and/or cement content on the one hand and the rate of erosion on the other hand, as can be seen from Fig. 59.

The resistance to abrasion of some concretes made with blast furnace cement has been determined by *H. Gibbels* [55] by means of a Böhm's wheel. The abrasion of the unfired material was in general slightly greater than that of samples which had been fired at 1200 °C. The resistance to abrasion drops to a minimum in the middle range of temperature in a way analogous to that of the structural strength.

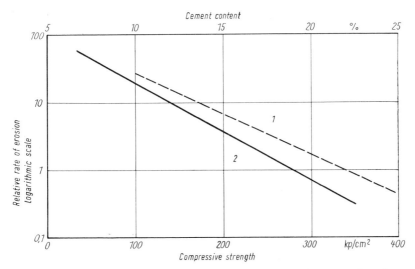

Fig. 59. *Relationship between cement content, compressive strength and rate of erosion for aluminous cement concretes (Wygant and Bulkley)*
1 Cement content; *2* Compressive strength

5.1.5. Chemical properties

In refractory science the description "chemical properties" is used to cover the behaviour of refractory materials in relation to other media, for example the reactivity of refractory substances amongst themselves or with other materials with which the furnace lining comes into contact, such as cement clinker, glass or slag. It also includes corrosion by smelted materials, dusts and gases, the ingress of foreign substances during operation and their diffusion through the material with consequential changes in material properties. Since refractory concretes do not usually come into direct contact with smelts, it is principally only the behaviour in relation to dust and ash deposits and the effects of gases, slags and other corrosive substances present in the furnace atmosphere which are of concern. Direct attack from alkaline, oxidic or metallic smelts seldom occur in practical operational use. The use of refractory concretes in nonferrous metallurgy has, however, recently increased, and therefore a start has been made in studying their behaviour in contact with molten metals.

The behaviour of a refractory material in the presence of corrosion and its resistance to it are a function of very many factors. Among the most important are the chemical relationships between the two partners, particularly the acid-basic relationship, their affinity, the magnitude of the free reaction enthalpies occurring during chemical changes, the porosity of the material and the temperature. Temperature in particular plays a decisive role, since the reactivities of all substances increase sharply with temperature. In addition, the viscosity of a smelted material decreases with temperature, while its wetness increases, leading to an acceleration of all exchange processes. High porosity, which means a larger surface exposed to chemical reactions, also leads to increased susceptibility to mechanical erosion.

In relation to reactivity, the chemical and mineralogical composition of concretes (that is to say, the degree of basicity or neutrality of the component materials) is of extreme importance. In any practical application this aspect must be considered together with the question of temperature of operation. It can be assumed as a fundamental rule, however, that refractory concretes made with suitable proportions of chrome ore and aluminous cement will behave more or less as chemically inert materials, as indeed do firebricks of similar composition [183].

It is very important, when considering fire-resistant concretes for use under chemically corrosive conditions, to possess knowledge of their behaviour in the presence of oxidising and metallic substances in solid or liquid form or in the presence of gaseous media. A number of typical problems will therefore now be briefly discussed.

5.1.5.1. Behaviour in the presence of slagging (scorifying) due to oxides and smelted materials containing oxygen

L. Ludera [184] has made a systematic series of investigations into the effects of coal slags on refractory concretes made with portland cement and a variety of different aggregates. Tests with three acidic slags having melting points between 1150 and 1270 °C indicated that refractory concretes made with chamotte, chrome ore or magnesia as the aggregate, and chamotte powder, clay, chrome ore dust or sintered magnesia dust as the stabiliser, possessed an extremely high degree of resistance The insignificant amount of attack which did occur was uniform, and the formation of cavities, so typical of slag attack, was absent.

Similar observations on the behaviour of coke oven doors (operating temperature 1000 °C) justify the assumption that refractory concretes are extremely resistant to attack from coal ash and coal slag This is confirmed by the fact that doors of this type last in service for about one year.

Interesting observations have been reported by *M. N. Kajbičeva, L. Ja. Pivnik* and *N. I. Mar'evič* [185] about the behaviour of high alumina cement concrete with corundum-chamotte, fused magnesia or chrome ore magnesia aggregates under the operating conditions which are found in electric furnace roofs and also about the chemical reactions which take place in the different temperature zones of the chamotte lining. In the fire zone FeO, Fe_2O_3, MgO and MnO become concentrated and lead, in the case of chamotte concrete, for instance, to the formation of $FeO \cdot Fe_2O_3$, $FeO \cdot Al_2O_3$, $MgO \cdot Al_2O_3$ and $MgO \cdot Fe_2O_3$.

The following basic rules are given by *E. Hammond* [165] for the chemical resistance of aluminous cement concretes: concretes made with chrome ore, chrome ore magnesia or magnesia possess a good resistance to basic slags; concretes made from corundum are especially resistant to both acid and basic slags. Sillimanite concretes behave in a similar manner; for example they have shown a high degree of resistance in practice to soda slags in desulphurating furnaces.

Resistance to acid slags can be improved still further by the use of suitable surface coatings. Some examples of these are mixtures of chrome ore and fire-resistant clay with sulphite liquor, chamotte and clay with waterglass or quartzite shale, fire-resistant clay and glass powder with waterglass. These have been successfully tested by *Ludera*.

5.1.5.2. Behaviour in the presence of molten metals

The effects of molten metals upon fire-resistant concretes have so far been investigated mainly in the context of non-ferrous metals.

Research by *L. Ludera* [184] with melted antimony and tin (in the temperature range 1200 to 1350 °C, duration 4 hours) has shown that chrome ore-chamotte concretes are relatively the most resistant of all refractory concretes made with portland cement, although even with these concretes fairly deep zones of diffusion were found. All other refractory concretes were badly corroded and possessed cavities, infiltrations and layers of diffusion.

According to information published by *D. F. Stock* and *J. L. Dolph* [186] concretes have been used with varying degrees of success in smelting furnaces in the aluminium industry. The principal difficulty lay in the low strength in the critical temperature range of 800 to 1100 °C. The authors were able to show by systematic tests, however, that a concrete made from cement of high alumina content and previously fired to 1100 °C exhibited no corrosion at all from molten aluminium and was the best refractory material for side wall linings under these conditions.

Fire-resistant concretes made with cement of high alumina content (A:C $\approx$ 2,5) such as the chamotte or corundum concretes described by *W. H. Gitzen*, *L. D. Hart* and *G. MacZura* [166] which contained 15 to 25% of cement, have proved to be resistant to soaking and penetration by molten metals.

5.1.5.3. Behaviour in the presence of gases

Refractory concretes under operating conditions are usually subjected to a wide range of gases. Their effects cannot be neglected and attention should certainly be paid to the exchange reactions between them and the concrete. There is, unfortunately, little available information on this subject.

The effect of gases containing sulphur (particularly SO_2) on concretes made with portland and aluminous cement has been the subject of investigations by *K. G. Bergman* [187]. This condition is encountered in electrostatic filters, to quote an example. It is surprising to note that in the case of portland cement concrete there is an increase of strength after heat treatment at 400 °C which is accompanied by an increase in the weight of the test pieces, indicating the

11*

formation of sulphates from the SO_2 in the gas. At 800 °C there is a pronounced decrease in the strength, as would be expected under normal circumstances.

The question of the behaviour of concrete in reducing atmospheres must also not be ignored, although precise information in this respect is so far lacking. There are indications, however, [165] that in the middle temperature range a reaction between the iron compounds in the cement and, for example, CO may be expected; in these cases iron-free cements should therefore be used if possible.

5.1.6.   Electrical properties

The electrical conductivity of a refractory concrete made from high alumina cement (75% Al_2O_3) and corundum chamotte (83% Al_2O_3) has been measured by *S. A. Žicharevič* and his co-workers [188], in connection with the insulation

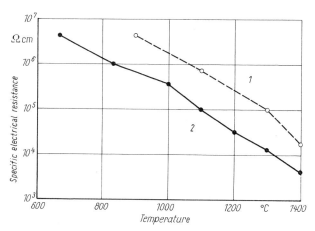

Fig. 60. Temperature dependence of specific electrical resistance of high alumina cement concrete (Žicharevič and colleagues)
1 unfired; *2* fired at 1650 °C

of electrode coolers for electric arc furnaces. The relationship between electrical resistance and temperature for both fired and unfired materials is shown in Fig. 60. The resistances are not very different from those of other high quality refractory bricks, such as sintered corundum, corundum chamotte, magnesia, silica and chrome ore magnesia.

5.1.7.   Special properties of lightweight refractory concretes

In the field of concrete technology materials have been evolved for the purpose of providing insulation at high temperatures, which are analogous to the well-known lightweight firebricks (the latter usually have a chamotte basis). These materials are known as lightweight refractory concretes. In contrast to normal concretes, which possess normal, average density, and in contrast therefore to normal high temperature dense concretes, these lightweight concretes possess certain properties which depart from the norm. These special properties are now briefly discussed.

Lightweight refractory concretes do not, however, differ in respect of all the properties so far discussed. In fact, dense and lightweight concretes possess many properties in common. The reactions and phase-building processes that take place at elevated temperatures, for example, can be regarded as basically the same, so long as the chemical compositions are the same. Stability under the effects of temperature are also of the same order. The temperatures at which softening under stress takes place differ, however, because the deformation of a material is a function not only of the softening, but also of the structural strength. The variation of strength with the temperature at which the concrete has been fired will also follow a similar pattern to that of the dense concrete; in other words there is a critical region in the middle temperature range, in which the strength drops to a minimum. A parallel can also be expected in the expansion/shrinkage behaviour, where the irreversible volumetric change of the lightweight concrete will presumably be rather larger than that of the dense concrete, since the phenomena of sintering at high temperatures are especially pronounced in the lightweight material.

Pronounced differences from normal refractory concretes exist in all those properties which are associated with the high porosity of lightweight concretes. The principal ones are: density, compressive strength, resistance to abrasion, stability under fluctuating temperature, permeability, thermal conductivity, stability under compressive stress at high temperature, and creep resistance. The reduced chemical resistance, which is typical of all materials of high porosity, is not important in this case, because insulating materials do not normally come into contact with aggressive liquid media. This property therefore does not need to be discussed here.

5.1.7.1. Density, porosity and water absorption

An especially noticeable feature of lightweight concretes is their high porosity. It is well known that this is achieved either by the use of porous aggregates (such as ceramsite or blast-furnace pumice) or by foaming of the wet concrete by means of foaming agents or the introduction of a gasified structure just before the onset of the reactions of setting.

The densities of lightweight concretes lie in the range 0,4 to 1,6 (25 to 100 lb/ft³) depending upon the process of manufacture. Concretes with densities between 1,4 and 1,6 (87 and 100 lb/ft³) do not really belong to the lightweight class. In the standards of the German Democratic Republic [173] they are described as concretes of medium density. True lightweight concretes, according to the standard WST 99-12-3 are made by the foaming process and possess densities not exceeding 1,2 (75 lb/ft³) independent of the type of cement. Soviet regulations prescribe that lightweight concretes made with portland cement and ceramsite shall have densities between 0,9 and 1,4 (56 and 87 lb/ft³) [8]. Foamed concretes, according to M. Ja. Krivickij [117], possess densities of 0,5 to 0,8 (31 and 50 lb/ft³).

While the porosities of dense refractory concretes lie between 20 and 30%, those of lightweight concretes are appreciably higher. $K. Martin$ [118] has measured porosities of some of the foamed concretes which he has made and found values of 27 to 33% for open porosity and 52 to 57% for gross porosity. Lightweight refractory concretes made to the standard WST 99-12-3, for example, are required to have a gross porosity of at least 52%. In general porosities of from 40 to 60% should be achieved for insulating concretes.

The water absorption of lightweight concretes is, by their nature, high. It is a function, first, of the density or open porosity, secondly of the method of manufacture (for example, concrete with lightweight aggregates, foamed or gas concrete, normally hardened or steam cured) and lastly of the temperature of firing. Figures for water absorption available so far relate only to foamed concrete. Thus $K. Martin$ quotes figures of 22 to 30% for the water absorption of foamed concretes made with aluminous cement. $M. Ja. Krivickij$ has carried out systematic investigations into the water absorptions of various foamed concretes made with portland cement and fly-ash, granulated slag, sand or broken brick. From these it is concluded that the water absorption decreases with increasing density, but increases with increasing temperature of firing (measured up to 800 °C) by about 27% in the case of normally cured concrete, and by about 20% in the case of concrete cured in an autoclave. The lightweight concretes which were cured in an autoclave possessed lower water absorption values throughout than those which were normally hardened.

5.1.7.2. Strength

Lightweight concretes, due to their high volume of pores, possess very low structural strengths and low resistance to abrasion. It is evident, therefore, that they should not be considered for use in load-bearing members. A knowledge of their strength is, however, important from the point of view of transportability or the ability to carry their own weight. The standards, therefore, usually specify minimum requirements.

According to $K. Martin$ [118] foamed lightweight refractory concretes can be produced, having cold compressive strengths in the unfired condition of 20 to 70 kp/cm² (284 to 994 psi). The relevant standard of the German Democratic Republic requires a minimum of 25 kp/cm² (355 psi) for concretes made from portland cement or smelted aluminous cement. Foamed concretes have been made in the U.S.S.R. with compressive strengths of from 7 to 36 kp/cm² (100 to 510 psi) [117]. The maximum compressive strengths reached with ceramsite and vermiculite concretes are reported as 50 kp/cm² (710 psi) for a density of 0,9 (56 lb/ft³), 100 kp/cm² (1420 psi) for a density of 1,2 (75 lb/ft³) and 250 kp/cm² (3550 psi) for a density of 1,4 (87 lb/ft³) [8], [112], [189]. Pearlite concretes made with aluminous cement have a strength of about 75 kp/cm² (1070 psi) for a density of around 1,0 (62 lb/ft³) [190].

The strength does, of course, vary with temperature when the concrete is first subjected to heat. Since the strengths of lightweight concretes are in any case low in the unfired state, it is naturally important to know how much further they decrease when the concrete passes through the critical temperature range and whether the residual strength provides sufficient load-bearing capacity.

Research by *Krivickij* [117] indicates that the compressive strength of portland cement foamed concretes decreases by about 15 to 40% after annealing at 700 °C; in this case also, steam-cured concretes are superior to those which are normally cured. The variation of compressive strength with temperature for ceramsite concretes of different densities is given in Fig. 61 from results obtained by *M. G. Maslennikova* [112]. Measurements by *Martin* [118] indicated that losses of strength of more than 50%, and on occasions as high as 70%, occurred when concrete was heated to 1100 °C, the final values lying between

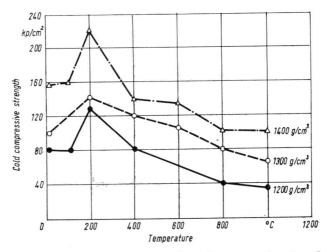

Fig. 61. Relationship between the cold compressive strength of ceramsite refractory concrete and its density and temperature of preheat (Maslennikova)

9 and 23 kp/cm² (128 and 326 psi). The value of cold compressive strength required for precast units of lightweight refractory concrete according to one (German Democratic Republic) standard [172] is a minimum of 8 kp/cm² (114 psi) after annealing at 1000 °C for 4 hours. Soviet regulations specify that ceramsite concretes should possess a residual compressive strength after heating to 800 °C of 40 to 50% of the original compressive strength [8]. A similar result can also be achieved with pearlite concrete (made with aluminous cement) according to *Ja. M. Gamarnik* [190].

The modulus of elasticity of pearlite concrete at 20 °C has a value of 100 000 to 130 000 kp/cm² (1 420 000 to 1 846 000 psi) and after firing at 800 °C about 30 000 kp/cm² (426 000 psi) [112].

5.1.7.3. Thermal-mechanical properties

The thermal-mechanical properties to which particular attention should be paid when making comparisons with dense concretes are the softening under pressure and the stability under fluctuating temperature.

As has already been pointed out, it is self-evident that the stability of a concrete under compressive stress at high temperature is primarily a function of the chemical and mineralogical composition of the concrete. In the case of materials of high porosity, however, there is also the purely mechanical effect of the concrete being compressed together as a result of the low compressive strength at high temperatures of the porous structure. This has proved to be important in normal insulating refractory bricks; it is therefore advisable to include a compressive test in the test schedule. This test should be at a lower stress than the standard one of 2 kp/cm^2; 0,5 kp/cm^2 (7 psi) would be a suitable figure [1; p. 260]. Technical Standard WST 99-12-3 attaches enough importance to this aspect to give figures for the behaviour of concretes under a stress of 1 kp/cm^2 (14 psi). For refractory materials possessing porosities equal to or greater than 50% TGL 13 713 requires a test loading in general of only 0,5 kp/cm^2 (7 psi).

Experiments to establish the softening under pressure of foamed lightweight refractory concretes containing smelted aluminous cement have been carried out by *K. Martin* [118]. The stability under compressive stress at high temperature decreases slightly with an increase in the cement content or in the size of the chamotte aggregate particles. Whereas pure foamed aluminous cement possesed a t_a-value of 1040 °C and a t_e-value of 1100 °C, the concretes exhibited t_a-values between 1080 and 1150 °C and t_e-values between 1170 and 1240 °C (the loading was 1 kp/cm^2 = 14 psi). The standard requires a t_a-value of 1100 °C and a t_e-value of 1220 °C for concretes made with smelted aluminous cement, and a t_a-value of 1050 °C and a t_e-value of 1150 °C for concretes made with portland cement. (For comparison purposes the t_a-value for a dense refractory concrete made with smelted aluminous cement is 1280 °C, and for one with portland cement, 1100 °C). For lightweight concretes made with portland cement and ceramsite, Soviet specifica tions prescribe a t_a-value (4% deformation) of 950 °C and a t_e-value of 1150 °C, when the density is 1,2 (15 lb/ft^3). The corresponding figures for a density of 1,4 (87 lb/ft^3) are 1000 and 1150 °C. For similar concretes *M. G. Maslennikova* gives t_a-values of around 950 to 1000 °C and t_e-values of 1170 to 1200 °C [112].

There is only a little information available about the stability under fluctuating temperature of lightweight refractory concretes. *M. Ja. Krivickij* [117] established the number of quenchings withstood by foamed portland cement concretes until the appearance of cracks. The samples were quenched in water from 800 °C and the numbers ranged from 3 to 13, depending upon the type of aggregate. The best results were obtained with fly-ash and quartz sand.

The same author has also studied the expansion of these concretes. This depends upon the density, typical examples for a lightweight concrete made with fly-ash being: for a density of 0,5 (32 lb/ft^3), 50 to 70 × 10^{-7}/°C; density of 0,65 (41 lb/ft^3), 60 to 80 × 10^{-7}/°C; density of 0,8 (50 lb/ft^3) × 10^{-7}/°C.

The permissible operating temperatures of insulating concretes are below 1000 °C. This limit is set on the one hand by the fire-resistance, and on the other hand by the practicability of insulating against heat.[1] Some information has been given by *M. Ja. Krivickij* [117] about the most suitable applications in operation, in relation to the type of aggregate and the method of curing. As a result of the relatively low operating temperatures, it is not to be expected

---

[1] According to *K. D. Nekrasov* [189] and *M. G. Maslennikova* [112] lightweight concretes made with portland cement can be used up to 900 °C, and those made with waterglass up to 800 °C.

that shrinkage much in excess of the normal will be experienced. Figures of 0,5 to 0,8% have been quoted for temperatures up to 600 °C [117].

5.1.7.4. Thermal condu ctivity

The thermal conductivity of lightweight concretes is of prime importance in connection with their use as insulating materials. A considerable amount of research has therefore been directed towards this property.

The thermal conductivity of this material depends to a great extent upon its density. This relationship was demonstrated experimentally for foamed concretes made with portland cement and fly-ash by *M. Ja. Krivickij* [117]. The values found, for a temperature of 250 °C, were: for a density of 0,65 (41 lb/ft^3), 0,06 kilocal/m hr °C; for a density of 0,8 (50 lb/ft^3), 0,10 kilocal/ m hr °C. Normally cured concrete possessed a slightly higher conductivity than concrete which had been cured in an autoclave. The standard WST 99-12-3 specifies that foamed lightweight refractory concretes made with portland cement, or with smelted aluminous cement and chamotte, and having a density of 1,2 (75 lb/ft^3) shall possess thermal conductivities not in excess of 0,27 to 0,30 kilocal/m hr °C.

The thermal conductivity of a large number of American lightweight concretes which had been pre-heated at 870 to 1090 °C have been measured by *W. C. Hansen* and *A. F. Livovich* [94], [191]. These concretes were made from smelted aluminous cement and lightweight aggregates, such as vermiculite, pearlite, expanded clay, etc. The higher the density, the higher, naturally, was the conductivity. The values obtained lay so close to a smooth curve that these authors recommend that the thermal conductivity can be obtained from the density by the use of the graph given in Fig. 62. Thermal conductivities at

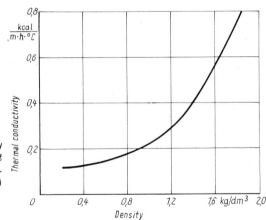

Fig. 62. Relationship between density and thermal conductivity for lightweight refractory concretes made with aluminous cement (Hansen and Livovich)

20 °C have been reported by *M. G. Maslennikova* [112] for ceramsite concretes made with portland cement and having densities in the range 1,3 to 1,4 (81 to 87 lb/ft^3), of about 0,32 kilocal/m hr °C.

Thermal conductivity increases with temperature, as it does in other materials for which $\lambda < 1$. The gradient varies, however, according to the aggregate and the method of manufacture, so that no universally applicable values can be given. The coefficients of thermal conductivity are still low enough at 800 °C, however, for the concrete to be an effective insulating material. The relationships between thermal conductivity and temperature for a selection of lightweight refractory concretes are shown graphically in Fig. 63.

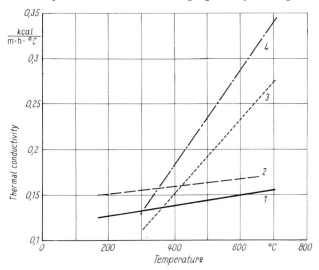

Fig. 63. *Variation of thermal conductivity with temperature for various lightweight refractory concretes (Hansen and Livovich, Krivickij)*

1 Aluminous cement : Vermiculite = 1 : 6; *2* Aluminous cement : Vermiculite = 1 : 4; *3* Portland cement foamed concrete, density 0,65; *4* Portland cement foamed concrete, density 0,8

Information has been published by *J. F. Wygant* and *M. S. Crowley* [192] about the thermal conductivity of refractory insulating concretes in gases which possess high thermal conductivities, such as hydrogen and helium. The gases may be met in the chemical and other industries.

5.2.   Refractory Concretes with Waterglass Binding Agents

The technology of high temperature concretes bound with waterglass and their use on a commercial scale in pyrotechnics have been developed to a considerable extent in the Soviet Union. Although materials of this type have been used in other countries from time to time, principally in the form of mortars, and have been described in some of the earlier publications on refractory science (for example, *A. B. Searle* [76; p. 735]) they were first developed by *K. D. Nekrasov* and his colleagues to the stage where from many points of view they are now of equal merit to the normal high temperature concretes and in some respects surpass them.

The discussion of the most important properties of waterglass concretes will therefore be based exclusively upon Soviet work.

5.2.1. Pyrochemical Processes

In discussing waterglass as a binding agent, it has already been stated that at relatively low temperatures water that was adsorbed and fixed in the silica gel is given off and at 600 °C crystallisation of $Na_2Si_2O_5$ takes place (see Section 2.2.1.3). It has been found in practical applications, however, that it is advantageous to add to the waterglass certain finely-ground substances, comparable to the "ceramic stabilisers" of portland cement refractory concretes, in the proportion of 20 to 25% of the weight of the concrete. Reacting quickly with the waterglass, these lead to the formation of a more fire-resistant binding agent between the coarse aggregates than the waterglass alone would provide.[1] These very fine aggregates serve as accelarators of the chemical reactions and, in certain circumstances, react with the waterglass even at room temperature, a phenomenon which has frequently been observed in practice. In addition they serve to lessen the tendency, a property of pure waterglass, to swelling under the effects of heat. Chamotte powder has proved to be a suitable additive for this purpose, in this case also. The use of commercial alumina should be avoided, since such a rapid formation of albite or nepheline takes place that the reactions leading to a proper binding are interfered with.

When the concrete is subjected to heat, further reactions occur between the binding agent and the aggregates; for example, combination with quartz to form polysilicates. In addition, when the temperatures are favourable, the crystallisation of quartz, crystobalite or tridymite from the dehydrated silica gel takes place. In contrast to the hydraulic binding agents, however, the pyrochemical processes in waterglass concretes, — except in the case of MgO aggregates — are not so thorough and penetrating; they do not, therefore, have such a marked effect upon the other properties, such as the strength. The research work carried out by *Ja. V. Ključarov* and *N. V. Mešalkina* [73] and *Ja. V. Ključarov* [193] has shown that in magnesia/waterglass concretes the formation of magnesium hydrosilicate, $MgO \cdot SiO_2 \cdot H_2O$, and perhaps also of $Na_2O \cdot MgO \cdot SiO_2 \cdot H_2O$, occurs first, followed, in certain conditions with regard to composition and temperature, by the formation of forsterite $2MgO \cdot SiO_2$, and the combining of $Na_2O \cdot MgO \cdot SiO_2$, and also, in the presence of chamotte, by cordierite, $2MgO \cdot 2Al_2O_3 \cdot 5SiO_2$. The formation of forsterite can be improved still further by the addition of quartz dust, leading to an improvement in thermal properties.

5.2.2. Irreversible Changes of Properties with Temperature

It is fundamental to the nature of all concrete-like fire-resistant materials, that changes of properties which are not reversible occur as a result of pyrochemical processes in the material. Changes of this type have already been described in detail in connection with the cement-bonded concretes. Similar effects also take place in waterglass concretes; they have not been investigated

---

[1] In the works of *K. D. Nekrasov* the solidified mixture of waterglass, sodium silicofluoride and very fine aggregate is therefore described as "cement brick" (*Translator's note:* literally "cement stone").

to the same extent, however, as in the case of the cement concretes and in addition are certainly not so strongly marked.

The group of properties under discussion includes strength, expansion/shrinkage behaviour, porosity and plasticity at high temperatures.

5.2.2.1. Strengths

Observations on hardened waterglass indicate that there is a loss of strength when it is heated, as far as 300 °C; the strength then remains approximately constant up to 900 °C. When fine and coarse aggregates are present the behaviour is essentially different: as the temperature is increased the strength also increases at first up to about 100 °C, then remains approximately constant and in the region of 600 to 700 °C varies again, the type of variation depending upon the aggregate. If the aggregates contain quartz there may be a loss of strength, while if chamotte is present an interim maximum value is reached, which can be attributed to the crystallisation of $Na_2Si_2O_5$. *K. D. Nekrasov* and *A. P. Tarasova* have made extensive investigations, which produced much fundamental information about heat-resistant waterglass concretes. This is described in their paper on the subject [42]. These authors made a series of experiments to ascertain the compressive strengths of these concretes in the

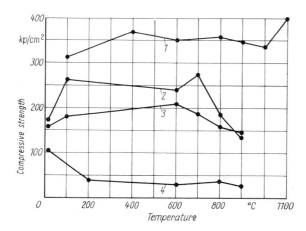

Fig. 64. *The hot compressive strengths of waterglass with very fine aggregates and waterglass/ chamotte concrete*

(Nekrasov and Tarasova)

1 Chamotte concrete; *2* Waterglass + chamotte; *3* Waterglass + chrome ore; *4* Waterglass + Na_2SiF_6

high temperature condition, in contrast to the more usual strength tests (at room temperature). Some typical strength curves underlining what has been said above are shown in Fig. 64. The effects described are, however, noticeable only in the case of pure "cement stone" or "cement brick" (see footnote, p. 171). In concrete itself, due to the small proportion of binding agent, they are of course much less marked, or may not be observable at all.

A noticeable feature of normal waterglass concretes is the more or less uniform value of the compressive strength over a wide range of temperature. In this respect they differ with advantage from the cement-bound refractory concretes. A most remarkable additional feature, demonstrated by *Nekrasov*, is that the

hot compressive strengths of samples (measured up to 500 °C) are higher than
the cold compressive strengths of samples which have undergone the same heat
treatment and of those which have not been subjected to any such treatment.
Still higher temperatures result in a further increase in the compressive strength
of samples tested after annealing, caused by the hardening of the concrete as
a result of the formation of smelted materials. This was especially noticeable
in concretes which had a silica base; in this case the compressive strength rose
at times to twice the initial value (at 110 °C) if the concrete had been annealed
at 1300 to 1400 °C [102].

More recent research by *N. V. Il'ina* and *L. I. Skoblo* [74] has shown that
this is generally true for a variety of concrete mixes based upon waterglass,
so long as they have been fired at a sufficiently high temperature. It can be
seen from Fig. 65 that there is a steep increase in the strength of chamotte
concretes above 800 °C which can on occasions reach values as high as three
times the strength of the concrete measured at 110 °C. In the case of magnesia
and chrome ore-magnesia concretes, however, the relationships are not so
simple; considerable fluctuations in the values occur and it is not possible to
lay down any firm rule.

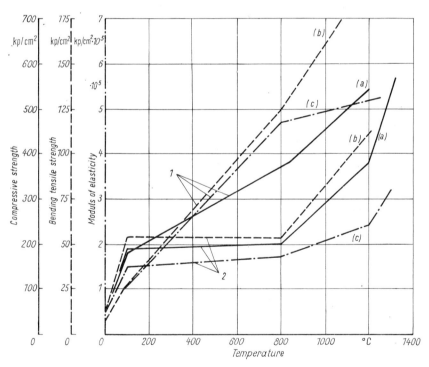

*Fig. 65. The relationship of compressive strength (a), bending tensile strength (b)
and modulus of elasticity (c) to the temperature of pre-heat, for waterglass refractory
concretes (Il'ina and Skoblo)*

1 Corundum chamotte; *2* Normal chamotte

The addition of Na_2SiF_6 has a profound effect upon strength. If the quantity is greater than 10 to 12% there is a steep drop in strength values above 800 °C. An increase in the waterglass modulus above 2,8 also has the effect of reducing the strength, at least in the range up to 800 °C. At higher temperatures, however, the lower alkali content of waterglass of this type takes effect, the proportion of smelted material being smaller and the fire-resistance consequently better. The particle size of the very fine aggregates does not appear to have much influence, it being acceptable to vary the fineness within fairly wide limits. It is, however, in any case preferable to use very fine material.

5.2.2.2. Expansion-shrinkage behaviour

The pyrochemical processes in heat-resistant concretes made with waterglass may also be expected to be accompanied by some dimensional variations when the concretes are first subjected to heat. Expansion/shrinkage behaviour of this type appears to a small, though noticeable, extent only when the mix contains a very fine aggregate. In the temperature range 100 to 400 °C an insignificant amount of shrinkage takes place, with a maximum value of 0,3%; at higher temperatures the normal thermal expansion occurs, followed finally above 900 °C by renewed shrinkage, resulting from the formation of smelted material and the onset of sintering.

The shrinkage effect in concretes at low temperatures can be quite small and may sometimes be neglected if the concrete has been dried at 110 °C. Both shrinkage and expansion vary according to the composition. In general, not more than 0,5% shrinkage need be expected to result from heating up to 1200 °C.

When the material is first heated, a slight volumetric increase sometimes occurs at about 700 °C. This is attributable to the crystallisation of $Na_2Si_2O_5$ and also depends to a great extent upon the composition of the concrete.

The phenomena just described are true for concretes other than silica concretes. They occur only the first time that the concrete is subjected to heat; any subsequent re-heating is accompanied by the normal reversible thermal expansion. Silica concretes exhibit a number of additional effects which are due to the presence of the high content of silica in the aggregate, such as volumetric step changes, irreversible growth due to the effects of the alkali, and so on. As a result of these processes the overall picture is rendered extremely complicated.

5.2.2.3. Porosity and permeability

Although only a small amount of information is available on the subject of variations in porosity, it may be assumed in general that the porosity passes through a maximum value at certain temperatures of firing and then decreases again at high temperatures. This has been demonstrated by *N. V. Il'ina* and *L. I. Skoblo* [74], for example in the case of chamotte concretes, as can be seen from Table 15. For silica/waterglass concretes, however, different results have been obtained [102], and it may be noted that practically no difference was observed between the porosities of the concrete in the unfired state and after firing at 1400 °C. In both cases the open porosity was about 23%.

Experiments to ascertain the permeability to gases of waterglass concretes made with chamottes and slags have been made by the VNII-Strojneft Institute [194]. The permeability in the unfired state is very low; it hardly varies at all up to 700 °C and does not increase appreciably until a temperature of 900 °C is reached.

Table 15. Variation of open porosity with temperature for chamotte waterglass concretes (Il'ina and Skoblo)

Temperature	Open porosity	
	Chamotte concrete	Corundum-chamotte concrete
[°C]	[%]	[%]
110	25,4	15,8
800	26,5	19,8
1200	23,2	19,6
1400	13,9	7,9

5.2.3.   Thermal-mechanical Properties

It is evident that the thermo-mechanical properties of waterglass concretes are of extreme importance in industrial applications, since they determine the permissible operating temperature level and the behaviour of the concrete at the temperatures met in practice. Much information is available from soviet sources on this subject, not only from laboratory work but for the most part from the results of practical operating experience.

5.2.3.1. Fire-resistance and softening under pressure

Information about the fire-resistance of waterglass concretes is relatively scarce. It is known from the work published by K. D. Nekrasov [6] that resistance to temperatures higher than 1700 °C has been achieved with mixtures of magnesia or chrome ore-magnesia and 20% waterglass. Temperatures of 1400 °C for chamotte with 25% of waterglass, and of about 1600 °C for silica with 18% waterglass, are also given. A fire-resistance of about 1900 °C may be expected for chromite concrete, according to L. A. Cejtlin [108], if the mix is correctly proportioned. A fire-resistance of 1800 °C has been established by N. V. Il'ina and L. I. Skoblo [74] for magnesia and chrome ore-magnesia waterglass concretes.

In comparison with the above temperatures, those for softening under pressure are much lower. For waterglass with additions of very fine chamotte or chrome ore aggregates (the material described by Nekrasov as "cement brick" or "cement stone") t_a-values of about 950 °C and t_e-values of about 1050 °C have been found, while the corresponding figures for magnesia and chrome ore-magnesia are approximately 1050 and 1580 °C. t_a-values as high as 1400 °C have been achieved by V. S. Sassa [195] under the right conditions by adding finely powdered chamotte or chromite to magnesia concretes. It should be possible to raise the temperature at which the concrete is still stable under compressive stress to 1660/1680 °C by the addition of quartz dust. These temperatures also depend to a great extent upon the composition of the waterglass, at least to the extent that a lowering of the waterglass modulus leads to a lower level of stability under compressive stress at high temperatures, due to the presence of a larger proportion of alkali leading to the formation of smelted material. A similar effect will be produced if the proportion of Na_2SiF_6 present is greater than 12% and the density of the waterglass is higher than 1,38.

Comparable effects are found in concrete-type mixtures, with the exception of silica concretes. In this case it is evident that the presence of the coarse aggregate fraction will have an additional influence, resulting from the strength of these aggregates. The t_a-values and t_e-values of a number of waterglass concretes, as given by various Soviet authors, are shown in Table 16. From this it can be seen that in general the thermal resistances are never lower than, for instance, those of refractory concretes made with normal commercial cements.[1] The wide range over which waterglass concretes of this type can be used can also be appreciated from this table; for mixtures containing chamotte these usually reach 900 to 1000 °C, and for some of the mixes 1200 °C and more. Special mixes for use in boiler plant at 1300 °C have also been described by *I. Ja. Zalkind* [197].

The table also shows that *P. N. D'jačkov* and his colleagues found that silica concretes possessed extremely good stability under compressive stress at high temperatures [102]. It is interesting to note that for these materials the t_a-values and t_e-values increased by 20 to 50 °C if the concretes were tested after firing at 1300 °C. A possible reason for this lies in the mineralisation effect of the alkali, which leads to a complete transformation of the SiO_2-phases.

5.2.3.2. Hot compressive strength, creep resistance and creep behaviour

Figures for hot compressive strength are known from the research carried out by *K. D. Nekrasov* and *A. P. Tarasova* [42]. This has also been discussed in Section 5.2.2.1 in relation to the changes which take place in strength under

Table 16. Temperatures at which softening under pressure takes place for waterglass refractory concretes

Very fine aggregate	Aggregate	t_a or 4% deformation temperature [°C]	t_e [°C]	Source
Chamotte	Chamotte	1040	1150	[42]
Chrome ore	Chrome ore	1100	1200	[42]
Chamotte	Chamotte	1100	1150	[149]
Magnesia	Chamotte	1220	1320	[42]
Magnesia	Magnesia	1280	1450	[149]
Magnesia	Chamotte	1320	1450	[196]
Corundum chamotte	Chrome ore magnesia	1370	1410	[74]
Corundum chamotte	Corundum chamotte	1450	1510	[74]
Quartzite	Silica	1560	1570	[102]

[1] Even higher values of t_a and t_e can be achieved by the use of suitable compositions [196].

the effects of temperature; reference should therefore be made to the results given in that section. Here it is desirable to emphasize once again the remarkably constant values of strength which obtain up to relatively high temperatures, a phenomenon which does not occur in the same way in other refractory concretes. Since the operating temperatures required for the concretes tested by these authors were not particularly high, the research extended only up to 900 °C.

Some values for creep at high temperatures are stated in the papers published by *K. D. Nekrasov* [6], *K. D. Nekrasov* and *A. P. Tarasova* [42], and *B. A. Al'tšuler, G. D. Salmanov* and *A. P. Tarasova* [168], in which these authors describe the results of their experiments.

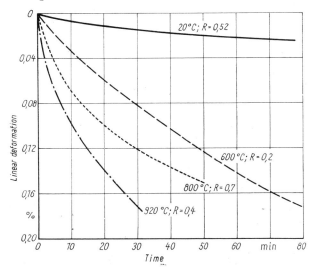

Fig. 66. The relationship of deformation to temperature, loading and time for waterglass chrome ore concrete (Al'tšuler, Salmanov and Tarasova)

The magnitude of loading R is the ratio of the pressure exerted to the maximum compressive stress at temperature T

The curves shown in Fig. 66 have been obtained from the work of the last-named authors and show the plastic deformation of a waterglass-chrome ore concrete in relation to loading, temperature and time. The factors related to breaking strength indicate the ratio of the loading used in the tests to the compressive strength of the concrete at the relevant temperature. It can be seen from this diagram that nominal values of plastic deformation occur from 600 °C upwards at relatively low loadings.

5.2.3.3. Stability under fluctuating temperature

Since waterglass concretes made with chamotte possess relatively low coefficients of thermal expansion (see Section 5.2.4.2), they can be expected to exhibit good stability under fluctuating temperature. Earlier experiments by

Nekrasov confirmed this fact; test pieces which had been quenched up to 10 times from 500 °C suffered practically no loss of strength. Later experiments under more severe conditions (quenching from 800 °C) provided renewed evidence of the outstandingly good stability of chamotte concretes under fluctuating temperature. The worst case found after 30 quenchings showed a maximum strength loss of 30% of the initial compressive strength.

Other authors have reported similar observations [74], [198]. The considerable extent to which this property depends upon the type of aggregate used is indicated, however, by the fact that the number of tolerable quenchings is considerably less in the case of magnesia concretes bound with waterglass. The number of temperature cycles is given in the published literature as only 5 to 8 [74], [193]. In the case of a chrome ore magnesia of a suitable grading the number may, however, be as high as 15. The stability under fluctuating temperature of silica-waterglass concretes is likewise influenced to a great extent by the mineralogical composition.

5.2.4. Physical and Mechanical Properties

There is not a large amount of information available about these properties from experiments under clearly defined conditions. Only a few quantitive values can be given here, therefore, of those properties that are of practical interest.

5.2.4.1. Density, water absorption, porosity

The densities and water absorption properties have been determined by *V. S. Sassa* [199] for hardened waterglass with various types of very fine aggregate (ratio in the range from 20:80 to 30:70). The values of density obtained when using different types of aggregate were as follows: chamotte $\approx 1,85$ to $1,9$, magnesia $\approx 2,0$ to $2,45$, chrome magnesia $\approx 2,6$, talcum $\approx 1,8$.

The water absorption figures for these materials are in the range 6 to 12%. The open porosities of waterglass refractory concretes differ very little from those of normal fire-resistant materials. Figures from a variety of publications indicate that in general values between 20 and 25% in the unfired condition may be expected. The porosity drops to about a half of the initial value after heating provided that the temperature reached is high enough.

5.2.4.2. Thermal expansion

When waterglass refractory concretes have once been subjected to heat, they exhibit the same normal reversible expansion as other fire-resistant materials. For the "dead-burnt" or over-burnt binding agent in this type of material (waterglass + Na_2SiF_6 + very fine aggregate) the values of the coefficient of thermal expansion have been determined by *Nekrasov* and *Tarasova* [42] for various types of fines. From the dilatometer curves up to 800 °C the following coefficients were obtained: for material containing chamotte, approximately $60 \times 10^{-7}/°C$; for material containing chromite, about $90 \times 10^{-7}/°C$; and for material containing andesite, about $70 \times 10^{-7}/°C$.

The same authors state values of between 60 and $80 \times 10^{-7}/°C$ for the thermal expansion coefficients of concretes made with chamotte aggregates, over the range 20 to 750 °C. The following values are given in the Soviet instructions for the use of heat-resistant concretes [8] (range 20 to 900 °C): magnesia concrete, $94 \times 10^{-7}/°C$; magnesia chamotte concrete, $56 \times 10^{-7}/°C$; concretes with chamotte, basalt, diabase or broken brick, $80 \times 10^{-7}/°C$.

5.2.5. Chemical Properties

One of the chief points of interest here is the question of the corrosion of concrete-like materials bound with waterglass by aggressive media at high temperatures. However, the possibility should not be ignored of aggressive condensates appearing in pyrotechnic plants when they cool down, which affect the concrete at low temperatures and thus influence its behaviour when it is again subjected to heat. This problem is principally one relating to the chemically-resistant, particularly acid-resistant, concretes. A considerable amount of research has been undertaken in this field into the behaviour of waterglass mortars and concretes in the presence of various media. This subject cannot, however, be discussed in detail here. Chemical exchange processes of this type are of interest, however, as already noted, in relation to the repeated subjection of the concrete to heat.

Experiments have been carried out by *Nekrasov* and *Tarasova* [42] to determine the effect of attack by water or steam upon the strength of the unfired material. It appears that the effect of water vapour (80 to 90 °C) or water is to lower the strength at temperatures up to 1000 °C only slightly. If, however, the compressive strength is tested in the presence of steam, the loss of strength in certain circumstances may be as high as 50%.[1] Other authors have also reported similar results [194].

V. S. Sassa [199] has communicated the results of extensive research into the stability of heat-resistant waterglass concretes in the presence of smelted sodium salts. It has been demonstrated that excellent resistance is possessed by concretes made with magnesia and chrome ore-magnesia in chemical plants in which smelted sodium sulphate, sodium sulphide, soda, sodium chloride and cryolite occur.

Further information about chemically-resistant refractory concretes made with waterglass may be obtained from the work of *A. P. Tarasova* [200] for temperatures up to 1000 °C. This author indicates that the resistance to attack of the concrete may be influenced by the use of special aggregates and compounds. Specially favourable results were obtained (except in the case of sulphuric acid) by the use of the material described as nepheline mud (a residue from the making of alumina, consisting essentially of C_2S) in combination with chamotte, chrome ore or talcum. A typical composition is stated to be as follows: waterglass 21%, nepheline mud 8%, chamotte powder 8%, chamotte sand 22%, coarse chamotte 41%.

---

[1] Resistance to the effects of steam in waterglass-chamotte concretes can be improved, according to *Nekrasov* and *Tarasova*, by the addition of small quantities of silica-like substances, as long as the operating temperature does not exceed 300 °C.

Aggregates of quartz, chamotte, or quartz-chamotte have proved satisfactory in acid gaseous or smelted media. Concretes made with magnesia, chrome ore or dunite are stable in the presence of smelted salts such as Na_2SO_4, NaCl, and NaF. Finally, in the presence of basic materials blast-furnace slag and nepheline mud are suitable aggregates for affording resistance to chemical attack.

The chemical changes which occurred in waterglass-chamotte concrete when used in an experimental rotating cement kiln have been studied by *N. V. Il'ina* and *O. G. Skoblo* [201]. The exchanges which occur between the clinker and the concrete are in principle similar to those with chamotte bricks; calcareous felspar occurs as a new phase. This results in some reduction of the fire-resistance, which, however, still reaches a value of 1650 °C.

5.2.6. Properties of Lightweight Concretes made with Waterglass

It is possible to make lightweight concretes with waterglass as the binding agent instead of portland cement. Ceramsite or expanded vermiculite may be used as aggregates, and chamotte as the very fine aggregate. Systematic research in this field has been carried out by *M. G. Maslennikova* in collaboration with *K. D. Nekrasov* and has been described by these authors [112], [202], [203].

For mixes composed of 22% vermiculite, 19% chamotte, 53% waterglass, 6% Na_2SiF_6, and of 48% ceramsite, 18% chamotte, 30% waterglass, 4% Na_2SiF_6, the densities lay between 0,75 and 0,85 and the strengths were 35 and 65 kp/cm² respectively (450 and 923 psi respectively.) The linear coefficient of thermal expansion is stated to be from 38 to $58 \times 10^{-7}/°C$ for the ceramsite concrete, and from 50 to $75 \times 10^{-7}/°C$ for the vermiculite concrete. The coefficients of thermal conductivity at room temperature were approximately 0,17 kilocal/m hr °C rising to approximately 0,38 kilocal/m hr °C at 600 °C. Despite the high porosity, the permeability to gases at low temperatures is very small; it does not increase to a great extent until the concrete is heated to 800 °C.

The maximum operating temperatures considered are 800 to 900 °C.

5.3. Fire-resistant Concretes Made with Magnesia Binding Agents

Although frequent reference is made in the earlier literature on refractory materials to the possibility of making concrete-like materials by using magnesia binders ("Sorel binders") and quite a large number of patents have been taken out for such materials (see references; for example *A. B. Searle* [76]) the technical development of magnesia-bound "concretes" and their commercial application did not effectively commence until relatively recently. In this case, once again, the basic work has been done in the Soviet Union, that of *A. A. Pirogov* being worthy of particular mention.

This class of concrete-like materials consists principally of sintered magnesia, sometimes with the addition of crushed chrome ore magnesia, and a magnesia binding agent. The constituents of the latter are finely powdered magnesia (caustic or sintered) or even finely powdered broken magnesia chrome ore bricks (approximate composition; 62% MgO, 11% Cr_2O_3, 5% SiO_2, 4% Al_2O_3 + TiO_2, 8% FeO, 4% CaO) and a solution of $MgCl_2$ or $MgSO_4$.[1] The mix proportions of the separate constituents will vary according to the purpose for which the material is required, but a ratio of coarse aggregate to binding agent of approximately 2:1 may be assumed. Detailed information about the more recently tested concrete mixes may be found in the publications of *V. A. Bron* and colleagues [204], *A. A. Pirogov* and colleagues [205] and *A. A. Pirogov* and *F. Z. Dolkart* [206].

5.3.1.   Irreversible Changes in Properties under the Effects of Heat

When concretes made with magnesia binding agents are first heated a number of chemical changes no doubt take place, especially as a result of the disintegration of the magnesium salts present in the binding agent[2]; the extent of these changes is not great, however, and certainly not so far-reaching as that of the transformations which occur in the case of cements. It is therefore not necessary to discuss them here. The formation of new phases as a result of the reaction between magnesia and the other aggregates, such as corundum or chrome ore, does not occur until very high temperatures are reached. Compounds of the spinel type then arise, in the range $MgO \cdot Al_2O_3$ to $MgO \cdot Cr_2O_3$ [208].

When, however, the decomposition of the so-called sorel binding takes place, there occurs a loss of strength which must not be ignored. At very high temperatures there is again an increase of strength, analogous to that found in the case of the cement-bonded concretes. This is due to the hardening which results from sintering. The magnitude of this increase in strength is a function of the "sinterability" of the material and therefore of its chemical and mineralogical composition. The loss of strength which occurs in the critical range is very large, as can be seen from the results of the work carried out by Soviet investigators; some typical curves are shown in Fig. 67 [204], [206]. If compressive tests are made on samples which have been heated to high temperatures, it is usually found that the strength is higher than that in the unfired state.

When the sorel binding changes to a sintered product, there is usually an associated change in the porosity of the material. It has been reported by *A. A. Pirogov* and *F. Z. Dolkart* [206] that the normally relatively low value of the open porosity (less than 20%) increases when the concrete is fired at 1100 °C to about 25%, but falls again to about 17% at 1500 °C. It is, of course, not possible to generalise about these figures, since each material behaves in a different manner. There are other observations which have indicated that the open porosities of concretes which have been fired at high temperatures are a few per cent higher than the initial porosities [205], [209]. In any event the level of temperature and the sinterability play the most decisive role.

The value of the permeability to gases of magnesia and magnesia chrome ore concretes made with sorel binding agents varies in a manner similar to that already described for concretes bound with cement and waterglass. Measure-

---

[1]) It should be pointed out here that magnesia binding agents are taken to include only those which have as a basis MgO + $MgSO_4$ and/or $MgCl_2$ or other soluble salts. Waterglass-bonded magnesia materials therefore do not belong to this group but to the waterglass concretes.

[2]) Other chlorides or sulphates, such as $FeSO_4$, may be used in place of magnesium salts [207].

ments made by *A. A. Pirogov* and *V. P. Rakina* [207] on mortars (30% sintered magnesia, 70% chromite, bound with approximately 20% of this quantity of $MgSO_4$ solution) indicated that the coefficients of permeability in the lower temperature range possessed negligibly small values. As in the case of other concretes, this coefficient started to increase at about 800 °C.

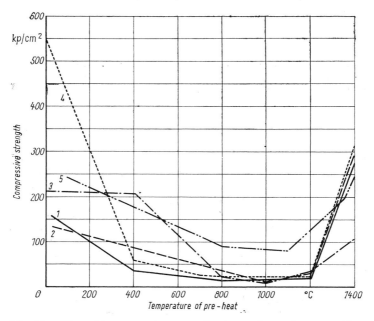

Fig. 67. The cold compressive strength after preheat of concrete-like refractory materials made with magnesia binding agents (Pirogov and Dolkart; Bron and colleagues)

1 MCr-concrete, 6% $MgCl_2$; *2* MCr-concrete, 6% $MgSO_4$; *3* MCr—MgO-concrete, 10% $MgSO_4$; *4* MCr—MgO-concrete, 10% $MgCl_2$; *5* MgO-concrete, 10% $MgCl_2$

Despite the variations in strength and porosity which occur with temperature, it should not be concluded that irreversible expansion-shrinkage behaviour is a feature of this material in the middle temperature range. Dilatometer measurements [206] have indicated that there is a fairly uniform linear expansion up to about 1300 °C. Above this temperature shrinkage sets in, in association with the commencement of sintering. This shrinkage may reach a value of more than 2% at 1500 °C.

5.3.2.   Thermal-mechanical Properties

The fire-resistance of the mixes considered is of a high order, on account of the constituents which have high smelting temperatures. Precise information about the cone fusion test points is therefore of no importance.

The resistance to compressive stress at high temperatures is of great practical importance since, as in the case of other refractory materials, it determines the permissible operating temperatures. These values, by the nature of the material, are also relatively high, being frequently in excess of the corresponding figures for cement and waterglass concretes. t_a-values of 1450 to 1600 °C and t_e-values of 1600 to 1700 °C can generally be expected. Precise data for particular mixes are given in Table 17.

Table 17.

Very fine aggregate	Aggregate	Binding agent	t_a or 4% deformation temperature [°C]	t_e [°C]	Author
Magnesia	Magnesia, Chrome ore	$MgSO_4$	1390···1480	1580···1590	[205]
Magnesia	Chrome ore magnesia	$MgSO_4$	1450···1550	1600···1700	[208]
Magnesia	Chrome ore	$MgSO_4$	1480	1540	[77]
Magnesia	Chrome ore, Titanium alumina Slag	$MgSO_4$	1450	1590	[209]
Magnesia	Magnesia, Chrome ore	$MgSO_4$	1520	1540	[241]
Magnesia	Dunite	$MgSO_4$	1580	1750	[208]
Chrome ore magnesia	Chrome ore magnesia	$MgCl_2$	1600	1610	[204]
Chrome ore magnesia	Chrome ore magneisa	$MgSO_4$	1640	1660	[204]

The resistance to compressive stress at high temperatures is dependent upon the previous thermal history of the material. Experiments have shown that for magnesia-chrome ore concretes, previous heating of the test pieces for 4 hours at 1600 to 1750 °C may result in an increase of the t_a-values by 130 to 200 °C and of the t_e-values by 20 to 100 °C [77], [205], [209].

Little information is available about the stability of this material under fluctuating temperature. *Pirogov* and his colleagues [205] report that the concretes developed by them withstand about 15 cycles of temperature when quenched from 1300 °C to room temperature in water.

5.3.3. Physical and Mechanical Properties

As a result of their composition concrete-like refractory materials bound with magnesia and having normal values of porosity possess densities in the range 2,8 to 3,0 (175 to 187 lb/ft³). If the material is fired, the density may fluctuate slightly in this range, due to the variation of porosity. If the proportion of chrome ore is rather high, the density may be slightly higher, up to about 3,3 (206 lb/ft^3).

Something has already been said about the porosity of materials of this type in Section 5.3.1. Generalised figures cannot be given, since the degree of compaction of the concrete will depend upon the methods used for working and handling it. The granulometry of the mix also has an effect. The values stated for the open porosity therefore vary somewhat. They lie between 14 and 18% for concretes in the unfired state, and between 11 and 23% for concretes which have been fired at high temperatures. In every case a more or less marked maximum will be found in the critical temperature range between these two extremes.

Concretes of pure magnesia content possess by their nature high coefficients of expansion. This is also a known feature of magnesia fire-bricks. Experiments by *A. A. Pirogov* and *F. Z. Dolkart* [206] showed a total expansion of about 1,7% at 1300 °C for mixes bound with MgCl$_2$. This corresponds to a linear coefficient of thermal expansion of about $130 \times 10^{-7}/°C$.

Due to the constituent materials a relatively high value of thermal conductivity is also to be expected. Thermal conductivities of about 1,5 kilocal/m hr °C at 880 °C have been reported by *A. A. Pirogov* [208] for chrome ore-magnesia concretes.

5.3.4.   Chemical Properties

Magnesia concretes frequently contain chrome ore or crushed chrome ore magnesia as aggregate or as very fine aggregate. As has already been stated in the discussion of these materials, the iron-chrome-spinel is very prone to reaction with the atmosphere. Possible variations in certain properties, which occur as the result of oxidising and reducing atmospheric conditions, may accordingly be viewed as the results of a change in the structure of the material caused by the change of valency of the iron. The same is also true for magnesia which contains a high proportion of Fe$_2$O$_3$. Experiments by *Pirogov* and *Dolkart* [206] certainly point to the fact that here also the atmosphere is not without nfluence.

Very little information is available about the behaviour of magnesia-bound concretes in the presence of aggressive media or other corrosive substances. Fundamental tests relating to corrosion do not yet appear to have been carried out. The only known facts are those from analytical investigations by *V. A. Bron* and his colleagues [204] of magnesia-chrome ore concrete blocks which had been built into a Siemens-Martin furnace; blocks made with MgSO$_4$ as the binding agent had absorbed less iron oxide than those made with a waterglass binding agent. It was concluded from this that the former type of concrete was more stable than the latter in the presence of attack by dusts and iron oxide vapour.

A. A. Pirogov, E. N. Leve and *P. D. Pjatikop* [210] have carried out investigations into the behaviour and chemical changes of magnesia and chrome ore-magnesia concretes during use in a Siemens-Martin furnace. It was found that attack by dust, smelted materials and vapours caused reactions with the basic concrete structure leading to the formation of magnesia spinels, which in turn formed solid solutions with one another (for example, between $MgO \cdot Cr_2O_3$, $MgO \cdot Al_2O_3$ and $MgO \cdot Fe_2O_3$) and so led to growth of the material. However, while smelted materials which were deposited or were formed in the surface usually diffused into the interior of the concrete block, iron oxides were concentrated mainly in the outermost reaction zones.

5.4. Concrete-like Refractory Materials Bound with Phosphate

Concrete-like materials made with phosphoric acid or aluminium phosphate as the binding agent have only recently come into use on a commercial scale. Therefore only relatively little information about their properties is available. The most important facts are briefly discussed here.

With regard to composition, it can be generally stated that in most cases Al_2O_3 is the aggregate used, both in the form of corundum and as normal commercial alumina or hydrate of alumina.[1] These are also the materials that have been the most thoroughly investigated with regard to properties and practical applications. The quantities of phosphoric acid and/or aluminium hydrogen phosphate solution are in the proportion of between 6 and 10% of the aggregate. In addition to alumina and corundum, other refractory materials, such as sillimanite, chamotte, and chrome ore [84], chrome ore-magnesia material [88] or chrome alumina slag [211] as well as silica [212] may be used. Special highly refractory compounds, such as ZrO_2 or BeO [189] and $ZrSiO_4$ or SiC [86] have also recently been used after binding with phosphate.

The efficacy of binding of the phosphoric acid or phosphate is a function of a number of factors. These include the degree of concentration of the acid and the quantity present, the accelerator used, the type of aggregate and the temperature. With regard to acid concentration, the optimum proportion is given as a 2:1 dilution of 85% H_3PO_4 with water [86]. Too high a concentration leads to the formation of unstable compounds when the material has been subjected to heat; these then begin to break down in the presence of air. If the acid solution is too weak, on the other hand, insufficient binding strength will be developed [216]. The setting process may be accelerated by the addition of approximately 1% ammonium fluoride. The workability of the mix may be improved by the addition of clay in the proportion of 10 to 15%; this also has the effect of improving the physical properties of the mix.

Phosphate-bound corundum concretes are described by *A. Cser* [213] as being divided into the following types:

> cold-setting concrete,
> warm-setting tamped concrete,
> concrete mix to be stored dry,
> sprayed concrete.

K. D. Nekrasov [216] states that in his experience the most suitable fire-resistant alumina constituent is special fused alumina. The following compositions are given by *H. D. Sheets, J. J. Buloff* and *W. H. Duckworth* [86] as prac-

---

[1]) Concrete-like materials of this composition are described in the U.S.A. as "phos-tabs" (abbreviated from "phosphate bonded tabular alumina").

tical ones using alumina and zircon compounds (weights per cent):

corundum	77	zirconia	54
hydrate of alumina	9	zircon sand	18
H_3PO_4 (85%)	9	zircon	18
(+ inhibitor)		aluminium phosphate	7
water	4	water	2
NH_4F	1	NH_4F	1

The following mix is recommended by *L. Kolb* [319] for tamped concrete (weights per cent):

	coarse	fine
fused alumina or corundum 20 20	20	—
fused alumina or corundum 120 20	20	20
fused alumina or corundum 220 40	40	—
fused alumina or corundum 280 .... —	—	60
hydrate of alumina 9	9	9
H_3PO_4, 83% 11	11	11

In the German Democratic Republic at the present time pressed concretes with a phosphoric acid binding agent are made in the VEB Chamotte and Clinker Works at Meissen.

5.4.1. Pyrochemical Processes

Some of the reactions which take place in mixes bound with phosphates have already been discussed in Section 2.2.4. This aspect of the subject will now be dealt with more fully here.

Mixes of the cold-setting type, which contain very reactive aggregate constituents, start to react after only a few minutes. The setting time can, however, be extended to a few hours by the use of suitable methods of cooling [213].

The use of warm-setting concrete mixes is, however, common and these may be stored for a fairly long time even after the binding agent has been added. Concretes of this type set very slowly at room temperature, a feature which is useful in relation to transport and storage, but is not favourable to the development of the structural strength. The processes necessary for the start of the chemical reactions and the development of the bond are not initiated until the temperature is raised; temperatures between 100 and 400 °C are necessary for this purpose. Even the cold-hardening types do not develop their full bond strength until a temperature of about 350 °C is reached.

It is essential to choose the correct sequence of heating, if the maximum strength of binding is to be developed in the material and the optimum properties ensured. A suitable procedure was found by *K. D. Nekrasov* to be as follows: first of all 8 hours at 110 °C, followed by 8 hours at 220 °C and finally 16 hours at 450 °C.

The mechanism of setting has as its basis a reaction of the phosphoric acid or acid phosphate with the aggregate, in which metaphosphates $[Al(PO_3)_3$ or others] are formed with complete neutralisation of the relevant orthophosphates $[AlPO_4, Mg_3(PO_4)_2, CrPO_4]$ or — particularly in the case of incomplete reaction — through a condensation process. The small quantities of commercial alumina and/or hydrate of alumina or of clay often added to the mix accelerate the combining process by their relatively high reactivity. Caustic magnesia operates in a similar manner but the proportion added must not exceed 1.5% [84]. Since the reactions may be presumed to pass through pyrophosphoric acid as an intermediate phase, pyrochemical processes may be said to have already arisen in this case during the setting process.

An increase of the temperature above that which produces the optimum value of bond strength leads to further reactions, including certainly that of $Al(PO_3)_3$ with Al_2O_3 leading to the formation of $AlPO_4$ [213]. This compound, which no doubt is the principal cause of the binding process, starts to break down at high temperatures accompanied by the separating out of corundum. This newly-formed corundum phase may be presumed to contribute further to the strength.

An undesirable phenomenon which can sometimes occur during the heat treatment or drying process is that of gaseous expansion [214]. This is attributed to the development of hydrogen, amongst other causes, which has been formed by reaction between the residue of metals and the acid; for this reason inhibitors are sometimes added to the mix [86]. This effect may, however, be used under certain conditions for the formation of insulating concretes of low density [88].

5.4.2. Irreversible Changes of Properties with Temperature

The variation of compressive strength with temperature follows a characteristic pattern. Starting at room temperature it increases in every case to a more or less marked extent, a phenomenon attributable to the development of chemical binding which has been referred to above. If phosphoric acid is used as the binding agent, there is a continuous increase of strength right up to high temperatures [211], [215]. With aluminium phosphate as the binding agent, however, a maximum value is reached at about 600 °C, followed by some loss of strength up to 900 °C and then a further increase [84]. Some typical curves of strength are shown in Fig. 68.

In contrast to concretes bound with cement, the materials now under discussion have the great advantage that either they do not possess a critical strength region of any significance or that their loss of strength is largely negligible. A certain disadvantage is the low initial strength of 10 to 50 kp/cm² (142 to 710 psi) at room temperature. The concrete must be heated to at least 100 °C to improve upon this.

The measurements which have so far been made of open porosities show that variations of this property with temperature do not appear to follow any law. The open porosities of concretes made, for example, with slag vary [211]; at 1600 °C a slight increase is detectable. Other mixes [215] show a sharp reduction in porosity in the high temperature region, the figure sometimes dropping almost to zero at 1600 °C. This drop is accompanied by a rise in the density from 2,4 to 3,5 (150 to 218 lb/ft³) in the case of corundum concretes. The pattern of variation will obviously depend upon the composition of the concrete.

The amount of shrinkage induced by high temperatures is in general negligible [82], [84], [211] or reaches at the most a value of 0,4% [181], [216].

5.4.3. Thermal-mechanical Properties

Phosphate-bound concrete-like materials exhibit a degree of stability under compressive stress at high temperatures comparable to that of normal refractory materials of similar composition. Thus the chamotte concretes have t_a-values of about 1300 °C and t_e-values of 1400 °C. The corresponding figures for mullite concretes are given as around 1400 and 1430 °C, while those for silica concretes lie in the region of 1480 and 1510 °C respectively. t_a-values of

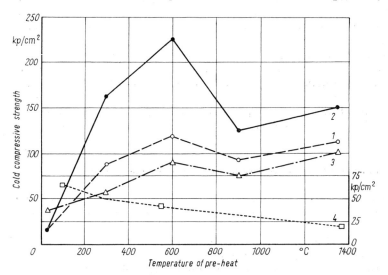

Fig. 68. Variation with the temperature of pre-heat, of cold compressive strength in refractory, concrete-type chamotte materials bound with phosphate (Bechtel and Ploss) and bending modulus in phosphate-bound Al_2O_3 (Sheets, Buloff and Duckworth)

1 Chamotte; *2* Chamotte + Al(OH)₃; *3* Chamotte + MgO; *4* Al₂O₃

1500 °C and t_e-values of 1600 °C may be reached with alumina [84]. For concretes made with chrome alumina slag, figures of 1470 and 1740 °C respectively have been reported [211]. *K. D. Nekrasov* [216] gives a t_a-value of about 1700 °C for phosphate-bound corundum.

It may be seen from the above figures that, depending upon the aggregate, fairly high operating temperatures may be achieved with phosphate concretes. Apart from the normal ranges, specific peak values, which permit these concretes to be used as highly refractory materials, are of particular interest. Thus experience in the Soviet Union has confirmed that corundum concretes can reach an operating limit of 1800 °C [305]; still further improvements may be expected. In the U.S.A. working temperatures as high as 1900 °C have been achieved [224].

The values of compressive strength under temperature are outstanding. *K. D. Nekrasov* and *A. P. Tarasova* [305] speak of a figure of 600 kp/cm² and are even said to have achieved 1600 kp/cm² (8500 and 22700 psi). In concretes made in America a high breaking strength in bending has also been demonstrated at high temperatures. It is interesting to note that the bending modulus, when the sample is tested in the hot state (up to 800 °C), is higher than that of the cooled sample, as illustrated in Fig. 68 [86].

The stability of these materials under fluctuating temperature is generally good. Tests in accordance with recognised standards showed that from 15 to 20 quenchings were sustained. Tamped concretes of this type have proved successful for tunnel furnace wagons, withstanding more than 100 entries into the furnace without visible evidence of damage [84].

5.4.4. Physical and Mechanical Properties

From what has been said in the foregoing Section it can be seen that refractory materials bound with phosphate and subsequently heated to temperatures adequate to ensure that the full bond of the phosphate is developed are capable of good strength properties both in compression and in the bending tensile test. This results not only in high resistance to static forces but also in a high abrasion resistance. This quality is retained even at high temperatures and results in the abrasion resistance of the material being about 40 times as high as that of normal fired bricks [222]. Materials of this class are particularly noted for their high resistance to erosion up to 1800 °C [82], [319].

The porosities of phosphate concretes vary and are dependent upon the method of manufacture. Values of open porosity for mixes which have set at temperatures of 100 and 400 °C respectively are known to lie between 20 and 30% [211], [215].

The density, being a function of the composition and the porosity, consequently varies in a similar manner. For corundum concretes, values of approximately 2,4 (150 lb/ft³) for cold-setting material and of 2,5 (156 lb/ft³) for hot-setting material have been ascertained [215].

Lightweight concretes of high refractory quality can be made from corundum mixes, according to *K. D. Nekrasov* and *A. P. Tarasova* [305]. They may be subjected to temperatures of at least 1500 °C and are already in extensive use, their cost being only half that of normal high-alumina materials.

5.4.5. Chemical Properties

Although it is not possible to find results of systematic laboratory research in this field, the successful use of phosphate bound concretes in many types of work is evidence of the high resistance of phosphate as a binder, particularly in the presence of normal flame gases. It has recently been reported from the Soviet Union [305] that concretes of this type also exhibit better resistance than any other refractory material to the attack of chloric gases at high temperatures. In addition the use in crude oil processing plants and in plants used for the extraction of titanium-chlorine compounds, of corundum concretes bound with phosphate is evidence of high resistance to chemical attack.

5.5.   Heat-resistant Reinforced Concrete

Reinforced concrete is normally used for structures only if the maximum working temperatures cannot exceed 200 °C. In a very few cases operating temperatures up to 350 °C have been permitted. For use at higher temperatures normal reinforced concrete is protected by refractory brickwork or by a special gray cast iron. In addition metal-framed structures or frames lined with refractory brickwork are usually resorted to for load-bearing structures in furnaces and similar plants.

In many cases heat-resistant reinforced concrete provides a simpler solution than refractory brickwork. Nevertheless the design of this type of concrete can be very difficult; under the effects of high temperatures and all conditions of loading the reinforcing steel must act with the concrete in a manner that can be relied upon. These conditions include those of raising to temperature, long periods of operation under static load and steady temperature and also fluctuating temperature and cooling down.

5.5.1.   Particular Aspects of the Steel-concrete Complex at Elevated Temperatures

The bond between steel and concrete is the principal factor necessary to ensure that the steel and concrete work together in a structure. In a reinforced concrete which is to be subjected to high temperatures, the long-term resistance of the steel at these temperatures is of prime importance, when determining the maximum permissible temperature, the value of the calculated stress, the plastic properties, the stability of the structure and its resistance to the formation of scale.

In this connection it is not possible here to go in detail into the properties of steels under the influence of temperature. Only a few essential aspects of the problem will be mentioned in order to gain a better understanding [302]. The most important criterion is evidently the strength behaviour of the steel and particularly its yield point. Above a certain temperature the deformation of the steel at constant load will increase continually; the material will thus begin to creep at stresses lower than the yield stress. Creep becomes definitely noticeable at temperatures higher than 350 °C. In addition the ductility of the steel also plays an important part and must be taken into consideration when estimating the ability to resist the effects of heat. Steels to be used in heat-resistant reinforced concrete should therefore possess a satisfactory degree of ductility. High resistance to corrosion is also a necessary property of a heat-resistant steel, since if uninterrupted scaling takes place the cross-sectional area of the reinforcement will become continually smaller, leading to increases in the stresses. The suitability of the steel with regard to its physical properties, such as linear expansion and thermal conductivity, must be determined in relation to its operating conditions. It must be remembered that the linear expansion of steel may rise to twice that of heat-resistant concretes and even more. The thermal conductivity of steel becomes less with an increase in the proportion of alloying substances present; the linear expansion of high alloy austenitic steels is greater than that of low-alloy steels and carbon steels.

In addition to the normal shrinkage that takes place during the hardening of a concrete, heat-resistant concretes also suffer additional shrinkage when first subjected to heat. The reduction in the volume which occurs during the first process causes a certain amount of stress in the reinforcement, since frictional forces prevent sliding. This is only partially true for the second process, however, because of the higher expansion of the steel under the effects of elevated temperatures already referred to. The friction force and bond between reinforcement and concrete must also increase. The tendency to slip can be reduced and thus the shear resistance can be increased to a limited extent by the presence of irregularities on the surface of the reinforcement and by the use of indented bars. In the limiting case, however, the steel can force apart the surrounding concrete as the temperature increases. In research work use is made of the complex determination of the bond stress, referred to the unit surface area of the reinforcement [302].

5.5.2. Behaviour of Heat-resistant Reinforced Concrete with Cement as the Binding Agent

As already explained, the bond between the reinforcement and the concrete is a decisive factor. Research by *A. F. Milovanov* [302] into this property in the case of portland cement concrete indicated the following behaviour. There was a noticeable increase in the bond strength up to temperatures of 250 °C. The known increase in the strength of the concrete in this range resulted in an increase in the bond between steel and concrete. The heat also caused the reinforcing steel to press against the concrete to a certain extent; this also increased the bonding forces. At temperatures between 275 and 325 °C the value of the bond strength is approximately equal to that at room temperature. Further increase in the temperature results in an appreciable decrease in the bonding forces, the large difference in temperature deformations between steel and concrete at 450 °C resulting in a loss of bond strength of from 60 to 70%. By contrast, the bond strength of indented reinforcing steel is significantly higher; up to 450 °C there is hardly any loss when compared to the value at room temperature. With further temperature increase, however, a noticeable loss of bond strength is found to occur. The variation with temperature of the bond strength between steel and concrete can be seen in Fig. 68a. The steels used in these tests were a plain (smooth) reinforcing steel with a yield point of 2900 to 3180 kp/cm² (41 000 to 45 000 psi) and a hot-rolled reinforcing steel with a yield point of 3650 to 4550 kp/cm² (52 000 to 65 000 psi).

The better performance of indented reinforcement obviously results from the increased surface bond. The use of this type of reinforcement is therefore an important factor in achieving adequate strength in heat-resistant reinforced concretes.

If the temperature of the reinforced concrete is further increased above 500 °C the reinforcement begins to creep and the stresses in the concrete therefore decrease.

When a reinforced element is cooled it at first shrinks in the same way as a concrete element. At temperatures below 500 °C creep no longer occurs in the reinforcement and the shrinkage is intensified.

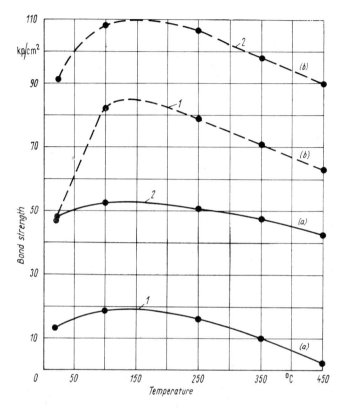

Fig. 68a. Variation with temperature of bond strength between reinforcing steel and heat-resistant portland cement/chamotte concrete [302]

1 Compressive strength = 185 kp/cm² (2630 psi) at 100 °C;   *2* Compressive strength = 240 kp/cm² (3400 psi) at 100 °C; (*a*) Smooth reinforcement; (*b*) Indented reinforcement

5.5.3.   Behaviour of Heat-resistant Reinforced Concrete with Waterglass as the Binding Agent

The structural strength of heat-resistant reinforced concretes with waterglass as the binding agent is likewise dependent to a decisive extent upon the bond between steel and concrete and its variation with temperature.

The compressive strength of concrete of this type decreases little or not at all with increasing temperature; therefore hardly any change takes place in the bond strength between steel and concrete. The bond of a smooth (plain)

reinforced bar has been found by *A. F. Milovanov* and *V. M. Prjadko* [303] to be as good at 275 °C as at room temperature. A further increase in the temperature caused a slight reduction in the bond strength, but at 450 °C its value was still 90% of the original.

When indented reinforcing bars are used, results as good as those obtained with portland cement concretes and described in Section 5.5.2 are achieved.

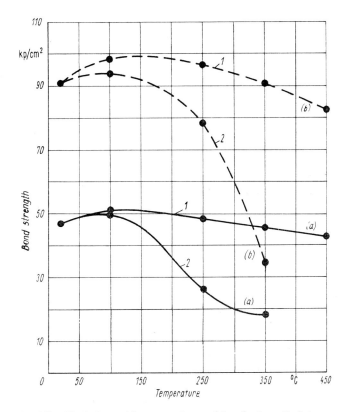

Fig. 68b. *Variation with temperature of bond strength between reinforcing steel and heat-resistant waterglass/chamotte concrete* [302]

1 Compressive strength = 365 kp/cm² (5180 psi) at 100 °C; *2* The same concrete after 10 temperature cycles; (*a*) Smooth reinforcement; (*b*) Indented reinforcement

The bond strength increases at first as the temperature is raised and has approximately the same value at 450 °C as at room temperature. Its variation with temperature can be seen in Fig. 68b. The same types of steel were used here as those in the experiments already described in Section 5.5.2.

One disadvantage of reinforced concrete made with waterglass binding agents is, however, that after a number of temperature cycles an appreciable loss of bond strength must be expected. A loss of between one-quarter and one-fifth of the strength was detected after 10 fluctuations between room

13 Petzold/Röhrs

temperature and 350 °C [303]. While the linear coefficients of thermal expansion of the concrete and the reinforcement differ by a factor of more than two, the thermal conductivities of steels are 30 to 35 times as great as those of concrete. The phenomenon known as temperature contact stress arises as a result of the deformation. The result of repeated heating and cooling is therefore that cracks appear and the bond between the reinforcement and the waterglass concrete is reduced.

No experimental results are so far known relating to the behaviour of this type of concrete at temperatures in excess of 500 °C. Because of the onset of creep in the steel, however, changes of properties may be expected similar to those which occur in heat-resistant reinforced concretes made with cement as the binding agent.

6. The Practical Use of High-temperature Concretes in Plant Operating at High Temperatures

In this chapter some typical examples of the use of high-temperature concretes will be described and the results of the experience gained will be set down. The summary is by no means complete. The aim, however, is to cover a wide range of practical applications, to derive experience from them and to indicate further possible uses.

The industrial use of high-temperature concretes varies in different industrialised countries, both in respect of the quality of the materials employed and also of the types of concrete used. Considerable efforts are being made in the Soviet Union, to develop these lining materials for more exacting duties [217], [218], [219], [306], [307]. Accurate statistics about the proportion of all refractory linings in use that are constructed of refractory concretes are unfortunately not available for all countries. The proportion of refractory concretes in the U.S.A. is quoted by a Soviet source (quoted by *G. G. Fel'dgandler* [220]) as being 9% of that of all other refractory materials. It may safely be stated, however, that the average overall figure is not more than a few per cent. Other examples of the use of concretes in plants operating at high temperatures come from France, Czechoslovakia, West Germany, Poland, England, Hungary, and Rumania as well as the U.S.A. The number of industrialised countries in which the use of refractory concretes is increasing today, is continually getting larger.

Until a few years ago the use of monolithic furnace linings was relatively uncommon in the German Democratic Republic and was confined to only a few plants. There has recently, however, been a marked change in this state of affairs and the trend towards the use of concrete for high temperature purposes is now noticeable in every branch of industry in which refractory plants operate. It is gratifying that, apart from the efforts that are being made on the part of the refractory industry, the firms specialising in the construction of furnaces are also becoming increasingly familiar with the new material. This is a natural consequence of its undoubted technical and economic advantages [221], [301], [308], [309].

6.1.   General Furnace Technology

Some of those general applications which occur to a greater or lesser extent in every other branch of the refractory industry will be discussed in this context.

One of the most important examples is that of chimney construction, in which the usual standard firebrick lining is replaced by concreting the walls with cast-in-situ concrete or lining them with prefabricated heat-resistant concrete segments. It is evident that this method of construction leads to a number of savings. Because of the relatively low temperatures, nomally lower than 600 °C, that occur in the linings of chimneys there is no need to use costly smelted aluminous cement; the cheaper portland or slag cements may be employed. Similarly, chamotte is not necessary for the aggregate, but instead basalt, or slags, to mention only two, are adequate, thus achieving further savings.

There is already experience in the German Democratic Republic of the use of heat-resistant concretes in chimney construction. Smelted aluminous cement, portland cement and chamotte have been used as constituent materials. Basalt and blast-furnace cement have also been experimented with.

Examples are also found of the successful application of heat-resistant concretes in chimney construction in other countries. An example described by *Ivanickij* [222] has as its constituents portland cement, bricks made from waste products, and sand, and was used, after being steam cured, for lining the flues of industrial furnaces and also the chimneys of block of flats. A concrete made from broken brick aggregate has also been used for the same purpose. Similarly it has been reported by *A. F. Milovanov* and *V. S. Zyrjanov* [223] that a satisfactory concrete for chimneys can be made using portland cement with aggregates of chamotte, basalt, diabase or andesite and that ceramsite concrete is also suitable. *K. D. Nekrasov* and *G. D. Salmanov* [214] have described the successful use in chimney flues of portland cement concrete made with tuff and (blast-furnace) pumice. Information about the design of chimney flues made from prefabricated heat-resistant concrete elements, for a number of different temperatures of the chimney gases, and about the composition of the concrete used, is contained in a paper by *A. F. Milovanov* [225].

Experimental investigations and calculations have been undertaken by *V. S. Zyrjanov* [226] in amplification of the practical experience gained in the construction and operational use of chimneys in which heat-resistant concretes were employed. The concretes were made from portland cement and chamotte aggregate and the investigations covered operating temperatures, crack formation and load-carrying capacity. Conclusions were drawn from these results about suitable mixes and design procedures for concretes in contact with flue-gases at temperatures of 500 and 800 °C.

A new step has recently been taken in the Soviet Union with the construction of complete chimneys from separate segments of heat-resistant reinforced concrete [224], [227]. The constituent materials used were portland cement, quartz powder as the stabiliser, and chamotte aggregate. The chimneys have a height of 30 to 40 m (100 to 133 ft) and a diameter of 1,25 m (about 4 ft 1 in) and were constructed from elements about 1,5 m (about 4 ft 10 in) long. Concrete rings constructed of two layers, the inner one being of heat-resistant lightweight concrete, have also been used. A waterglass concrete is often used for this inner layer, in order to provide high resistance to corrosion, the lightweight aggregates being either ceramsite or pearlite. Detailed information about heat-resistant pearlite concrete is given in a paper by *I. L. Majzel'* and *M. F. Sucharev* [310].

The chimneys are first assembled horizontally on the ground and prestressed, and are then erected by means of an erection mast. Fig. 68c shows a typical example of such a chimney, which in this case is 40 m (133 ft) high (quoted from reference [310]).

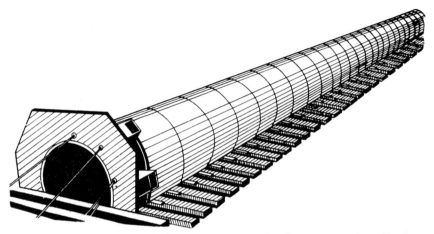

Fig. 68c. Chimney constructed of precast, prestressed, refractory concrete units shown pre-assembled before erection [310]

Detailed information about the design, construction procedures and erection methods used for prefabricated prestressed chimneys is given in the publications of *E. S. Vasil'ev* [311], *V. P. Vorslov* [312] and *A. P. Agurin* [313].

Experience gained in the Soviet Union [223], [227] indicates that there are definite advantages to be obtained by using chimneys made from refractory concretes, when compared with the more common materials such as steel and normal refractory products. These advantages exist irrespective of the temperature of the gas and of the dimensions of the chimney.

It must not be forgotten, however, that in the case of portland cement concretes the presence of acid gases can lead to corrosion as a result of the chemical reactions set up. This problem is particularly severe if the temperature of the gas is lower than the dew-point, so that acids such as H_2SO_4, HNO_3, HCl, are present in the liquid state. In such cases some protection against corrosion is essential. It has been demonstrated from experience gained in U.S.S.R. that coatings such as waterglass are not adequate. A waterglass concrete as an inner insulating layer has, however, proved effective. The question of the endurance of factory chimneys constructed of reinforced concrete has been thoroughly studied by *B. D. Trinker* [314].

There are official regulations in force in the Soviet Union relating to the protective measures to be taken against corrosion in flues and similar structures in which the surface comes into contact with aggressive media [300]. These regulations also contain condensed information about the protection of concrete and reinforced concrete.

Linings of refractory concrete have also proved their efficacy in other components of refractory plants which come into contact with hot waste gases, such as ducts for exhaust and other waste gases [107], [128]. Ducts of this type are made from precast components. A standard exists in the German Democratic Republic for the slabs and other segments used for this purpose [140]. This standard also covers gate-valves for use in exhaust gas ducts for operating temperatures up to 1200 °C.

Similarly satisfactory results appear to have been achieved with the use of portland cement-chamotte concretes for lining the roofs of producer gas furnaces [228]. The quantity of material required is less and the endurance of the lining is better than if normal chamotte brick is used.

The use of refractory concrete in heat exchangers and the experience that exists in this field has been described by A. E. Issers [227].

Fire-resistant concrete made with smelted aluminous cement has proved to be an extremely effective material for constructing the linings and fire-screens of locomotive fire-boxes, instead of the chamotte firebrick that has so far been the usual material for this purpose [229]. The effective life of the structure of 6 to 8 weeks has been extended by this change to more than 8 months. Equally satisfactory results have been obtained by the French Railways [122], the endurance of the concrete proving to be about three times that of firebrick.

6.2.   Iron and Steel Metallurgy

Ferrous metallurgy offers opportunities for a great variety of applications for high temperature concretes. Although the use of concretes was at first limited only to linings and similar structures operating at relatively low temperatures, with the continuing development and improvement of concretes of refractory and even highly refractory quality they have recently been used to an increasing extent for structures which are subjected to very high temperatures. Especially in those locations where the concrete is subjected only to relatively light wear, and where repairs during normal maintenance would consume an excessive amount of time, precast fire-resistant concrete or cast-in-situ fire-resistant concrete has been used with considerable success. There are many examples of this (see, for example, [229]).

The use of heat-resistant or fire-resistant concrete has been introduced to a greater or lesser extent in every branch of iron and steel metallurgy. The success of the results obtained has proved to be notable and has encouraged further experimentation under new operating conditions. From the experience gained so far about the properties of concretes, it is now possible to choose the necessary parameters for a concrete suitable for a particular duty.

The journal "Ogneupory" [230] provides a detailed reference work to technical literature, especially from Soviet sources, on the use of heat-resistant and fire-resistant concretes in the iron and steel industry. It covers early papers on the subject as well as lists of unfired plastic materials which do not necessarily belong to the true concretes.

6.2.1. Steelworks

Refractory concretes appear to have been first used in steelworks in the early nineteen-forties. Repairs to the boshes of blast furnaces were carried out in the U.S.A. in 1942 using aluminous cement chamotte concrete. Gas flues were subsequently lined with refractory concrete. The use of refractory concretes was introduced a little later in the Soviet Union, the first application being in the construction of blast-furnace foundations.

The use of portland cement concrete in the Soviet Union for foundations of this type dates back to the year 1949 [41], [231]. Practical experience gained from the use of a furnace floor slab constructed of refractory concrete over-lying the substructure has indicated that a thickness of about 30 cm (1 ft) is adequate to ensure protection of the furnace base even under extremely corrosive conditions and to limit the temperature of the upper side of the reinforced concrete foundations in the worst case to 250 °C. The constituent materials per cubic metre of this concrete were as follows: 300 kg of portland cement, 300 kg of finely ground chamotte as the stabiliser, 500 kg of chamotte sand and 700 kg of coarse chamotte, (504, 504, 840, and 1175 lb/yd³ respectively) the latter having a minimum compressive strength of 150 kg/cm² (2130 psi). Floor slabs have proved satisfactory in a large number of blast-furnaces. Their behaviour under operating conditions has been described by *Ja. M. Nemirovskij* and *B. A. Al'tšuler* [232].

Similar examples can be found in Britain and in the People's Republic of China [224]. Refractory concrete has also been used for the construction of the charging galleries of blast furnaces, for furnace gas extract flues and for the lining of hot gas ducts in blast furnaces. Refractory concretes made with aluminous cement are frequently used in the U.S.A. for repairing and effecting improvements to blast furnaces [233], [234]. The use of an almost completely iron-free cement is essential, in order to avoid attack in the presence of the reducing atmosphere. Further examples of the use of refractory concretes are found in the insulation of the body of the furnace and in the filling material behind the fire-resistant lining.

Chamotte concretes made with aluminous cement have proved very successful in the Soviet Union in blast-furnace hot blasts plants. Both rammed concrete and precast blocks have been used for this purpose [6], [224], [235]. Excellent results have been obtained with portland cement-chamotte concrete in the Kuznesk Steelworks [315]. Admittedly only small precast units of 0,6 ton weight were used, these being for the lining of the hot blast plant to a height of 28 m (≈ 92 ft). The economic advantages accrued from a reduction of the cost of the lining, a reduction of the heat losses, the achievement of high blast temperatures and a simplification of operating conditions. This hot blast plant has been in operation since 1958 [316]. More recently, much larger concrete units have been employed; the use of massive blocks up to a weight of 9 tons has been reported [236]. Approximately one half of the high-quality chamotte required in the lining was replaced in this way by concrete, the resulting economies being of a high order. More detailed information about the use of precast refractory concrete for the linings of hot blast installations can be obtained from the publications of *V. I. Bel'skij* [237] and of *N. D. Kiričenko* and *E. V. Kočinev* [238].

Refractory concrete made with aluminous cement is reported [98] to possess advantages when used in the construction of the lids of hot metal ladles, particularly from the point of view of durability. The ladles themselves may also be lined with rammed fire-resistant concrete; some examples, which resulted in appreciable savings in money and time, have been described by *R. Barta* and *B. Mokry* [317]. Similar economies are described by these authors arising from the use of concrete for the feeding heads of ingot moulds; although the durability of these was not increased when they were made from aluminous cement concrete, their repairs could be effected much more quickly than was possible if profiled chamotte bricks were used. Similar results have been observed in the case of the linings of the entrances to pig iron mixers; a cost saving of 75% was achieved by the use of aluminous cement-chamotte concrete instead of profiled bricks.

In the vacuum production of steel, fire-resistant concrete made from chrome alumina slag and a waterglass binding agent has been successfully utilised for the construction of the so-called "spray restrictors" of ingot moulds. The ingots produced in this plant are of a high standard [211].

For certain parts of the linings of acid converters the use of rapid-hardening concretes made with waterglass and cryolite is recorded [239]. Excellent results have also been reported of the use in the Soviet Union of a special binding agent known as VCZ (magnesia, iron filings and $MgSO_4$) for cold repair work to acid blast converters [240].

Fire-resistant concretes made with portland cement or aluminous cement have recently been used in various structures in Siemens-Martin furnaces, following the construction in the U.S.A. in 1943 of extraction ducts, for operating temperatures of 700 to 1000 °C, from smelted aluminous cement concrete. In this context the use is reported from the Soviet Union [41], [93] of portland cement concretes for certain lining components and lightly loaded structural members for temperatures between 500 and 1200 °C in regenerative furnaces and slag chambers. In addition to the use of dense concrete made with portland or aluminous cement (and chamotte aggregate) or waterglass (and silica aggregate) heat-resistant gas concretes made with portland cement and chamotte have also recently been employed in the construction of regenerators [318]. A study of the wearing properties of portland cement-chrome ore concrete in the flues and other parts of Siemens-Martin furnaces indicates that they are as resistant as normal refractory materials [241]. Considerable advantages have been found in the use of magnesia-chrome ore concrete (30% MgO, 70% chromite) with a sulphate binding agent instead of chamotte brick for the construction of the closures of charging openings. The normal life of 5 to 6 days could be increased to 35 to 50 days at a maximum [315].

Excellent results have also been obtained with the use of fire-resistant concretes made with chrome ore magnesia and waterglass as the binding agent in the lining of forked casting gutters in Siemens-Martin furnaces [242]. With the chamotte linings previously used the number of casting operations which could be tolerated was between 15 and 30 in the case of electric furnaces and only from 2 to 5 in the case of Siemens-Martin furnaces. When the above concrete was used, however, these numbers rose to between 540 and 600 in the first case and to between 40 and 60, on occasions even 170 to 190, in the second case.

Refractory concrete also appears to be excellent for the linings of casting pits; in this case only a small amount of chamotte is required as aggregate [98].

In electrical steel smelting it has been a frequent practice for some time to manufacture the covers of electric furnaces from fire-resistant concrete made with normal commercial aluminous cement. There are accounts in the relevant technical literature of the use of fire-resistant concrete with success in Hungary [214], the U.S.A. [243], and the U.S.S.R. for the insulation of electrode coolers and for the covers of electric furnaces [188], [244]. Normal refractory materials became excessively slagged during operation of the plant, leading to increased electrical conductivity and so to short-circuiting. The performance of concrete was better. The use of high alumina cement (basis CA_2) is essential, because the lower the lime content of the cement, the smaller is the degree of shrinkage of the concrete. Chamotte, in the ratio of four parts to one part of cement, or high alumina or high magnesia materials were used as the aggregate [185]. A well proved fire-resistant concrete made from high alumina cement (70% Al_2O_3) and corundum chamotte (82 to 86% Al_2O_3) in the ratio 20:80 [188] and possessing a fire-resistance of 1850 °C has been used in the Elektrostal steelworks and has been found to improve the durability of the furnace roof lining by 12 to 15% [245]. In American steelworks it has been shown that the use of concrete made with high alumina cement for the lining of electric furnace covers results in an increase in the number of permissible smelting processes to about 100 from the 45 to 60 uses previously obtained when silica was employed for this purpose. Blocks weighing 1,7 tons made from magnesia concrete have recently been used in the U.S.S.R. in electric steel furnaces used for the manufacture of manganese steel [205]. Magnesia-chrome ore concretes having $MgSO_4$ as the binding agent have been successfully used for the same purpose. They are not inferior to normal magnesia brick in any respect [204].

Pneumatic (gunned) concrete appears to have been used in the American iron and steel industry as an alternative to the more normal pouring process, particularly when carrying out repairs. The procedure of spraying is discussed in a number of papers, particular attention being paid to the differences between this process and that of casting the concrete [321], [322], [323]. Repairs to a blast furnace using the sprayed concrete method may require, for instance, only 8 hours, while the renewal of a refractory lining would take 3 weeks. The latter procedure is therefore far more expensive than concreting. In addition to these economic advantages there are also technical ones, in that there is an increase in the density of from 1 to 5%, in the modulus of rupture of about 50%, and in the cold compressive strength of up to 100%, by comparison with rammed or poured concrete. It has also proved possible to apply insulating concrete, such as pearlite concrete, satisfactorily by this process.

6.2.2. Forges and Rolling Mills

Fire-resistant concretes made with aluminous cement and chamotte were successfully tested in the U.S.S.R. as long as 20 years ago in a number of different metal working plants. Mention is made in a report by E. F. Gojkolov and A. V. Zotov [246] of the use of smelted aluminous cement concrete for lining a furnace forge (other than the roof) which had an operating temperature of 1400 °C maximum.

Pilot plants using smelted aluminous cement concrete had already been constructed in 1942 for heating and annealing furnaces [247]. In this case chrome ore concrete was used for the first time for certain parts of the furnaces in addition to the normal chamotte concrete. Since that time refractory concrete has been extended to a number of different plants in the Soviet steelworking industry. Satisfactory results are reported from the lining of soaking pit furnaces and pusher type heating furnaces with portland cement-chamotte concrete [248], [249], [250]. Good results have been obtained with the use of portland cement-chrome ore concrete having a t_a-value of 1500 °C in furnaces with operating temperatures in the range 1300 to 1350 °C [171]. The performance of this refractory concrete in the presence of mechanical abrasion and slagging from mill scale was found to be an improvement upon that of the chamotte lining previously employed. Its lower permeability to gases is another advantage.

In the German Democratic Republic refractory concrete has been used with success in soaking pit furnaces in the VEB Stainless Steel Works at Freital. Arches made in aluminous cement concrete possessed a working life three times longer than that of refractory brickwork. Even better improvements were found when concrete was used for covers, components for burners and parts of the walls. In these cases the improvement was four- or five-fold [229]. Similar improvements were demonstrated in annealing furnaces and other types of furnace at Freital and in various furnaces in the rolling mills at Burg. Information is given in a paper by E. Kuntzsch [325] about the latest developments in the partial mechanisation of the construction of industrial furnaces which have taken place in these plants using large precast units of refractory concrete.

Successful results have also been achieved with the use of waterglass concretes having a silica basis in the construction of soaking pit furnaces [251]. This concrete contained 62% of broken silica, 20% of finely ground silica material or quartzite and 18% of waterglass. This concrete proved to be of equal quality in every respect to normal silica brick at operating temperatures of 1400 to 1450 °C. The work referred to also provides interesting information about the wearing properties and changes in the chemical composition of the material in the relevant temperature zones.

In other countries there is also an increasing tendency towards the use of fire-resistant concretes for the linings of soaking pit furnaces [239].

The use of heat-resistant reinforced concrete slabs, made with portland cement, in annealing furnaces operating at temperatures in the region of 1250 °C has been described by P. I. Tolkačev and V. P. Borisov [252]. Smelted aluminous concrete also has been used with success in other countries for this purpose [98], [214], [253].

Chamotte concrete made with pozzolanic cement has proved to be satisfactory for insulating purposes in pusher type heating furnaces [157].

6.2.3. Foundries

The introduction of the use of refractory concrete to the iron-founding industry has so far progressed only slowly. It has proved to be a particularly useful material for concrete structures in the foundry buildings, rather than in the plant itself [214]. The very much higher resistance which it possesses to thermal changes, in comparison with normal concrete, makes it a suitable material for the construction of the foundry floor, which may have to withstand the spillage of molten iron. Normal concrete cracks, spalls and disintegrates under this type of treatment and so becomes quite unserviceable. Provided that attention is paid in the design and construction of the concrete to the points that ensure the material's resistance to fluctuating temperatures, it will prove quite satisfactory in service under the conditions referred to.

The introduction of refractory concretes into the construction of the foundry plant itself is taking place only very gradually. The remarks made in Section 6.2.1 about the uses of these concretes in connection with the casting processes in blast furnaces apply only to a limited extent to their utilisation in the founding process. Some practical applications have been described by E. Hammond [165]. The introduction of refractory concretes to this section of industry appears to be attended with considerable difficulties.

The use of blocks composed of chrome alumina slag in the mouth openings of furnace cupolas has been described by V. A. Bron and K. P. Semavina [211]. These blocks, however, had to be pre-fired and therefore do not belong to the true concretes. Unfired precast units and rammed concrete components may, however, be used in locations in which there is no attack from slags.

6.2.4. Other Metal Working Processes

The use of reinforced, high-temperature concrete, having as its constituents portland cement, finely ground chamotte and coarse chamotte in the proportions 1:1:3,5, for the linings to the covers of tempering furnaces is described by B. A. Al'tšuler and S. O. Rabinovič [254]. The operating temperature of the inner surface is 1050 °C, although the annular circumferential reinforcement does not reach a temperature higher than 350 °C. This concrete exhibited excellent stability under the extremes of fluctuating temperature to which it was subjected. The lining underwent more than 10 000 hours of service during two years of operation, withstanding 1850 temperature cycles due to the filling and emptying of the furnace, as well as 80 complete shut-downs [255].

The use of chrome alumina concrete with a waterglass binding agent for the construction of the burners of reheating furnaces was found by V. A. Bron and K. P. Semavina [211] to result in a threefold increase in durability by comparison with portland cement-chamotte concrete. A similar change, in this case from chamotte, resulted in the life of the gutters in induction furnaces being increased seven- or eight-fold.

Large-sized blocks of portland cement-chrome ore concrete have been successfully used for the linings of reheating furnaces for spring steels, with operating temperatures in the range 1250 to 1300 °C, in one Moscow work [216]. As a result of the high resistance of the concrete to thermal cycling, it was found possible to reduce the periods necessary for cooling down and reheating

13 a*

by a considerable amount, thus increasing the efficiency of the furnaces. Whilst a chamotte lining was found to require renewal every two months, the life of a concrete lining was proved to be at least 7 months.

6.3.   Plants Manufacturing Non-ferrous Metals

Among the non-ferrous metal industries, it is particularly in the aluminium sector that the possibilities of fire-resistant concretes have been exploited. A number of aluminium furnaces have been constructed using refractory concretes in the People's Republic of China (K. D. Nekrasov [256]). Crucible furnaces for the re-smelting of aluminium have also been constructed of fire-resistant concrete in the Soviet Union [196]. The mixture used comprised waterglass, coarse chamotte, fine chamotte, and finely ground magnesia in the proportions of 300:800:500:550 kg per cubic metre (504:1345:840:925 lb/yd³) of concrete. These furnaces have proved very satisfactory in service, withstanding $2^1/_2$ months of operation with up to 15 smeltings per day. The same material is also suitable for crucibles for other molten metals having temperatures up to 1200 °C.

In England fire-resistant concrete has been used in the construction of component parts of a large number of furnaces in the non-ferrous metal industries. Examples are linings, cover plates, gutters and burner-bricks [214].

6.4.   The Ceramics Industry

Refractory concretes were used at a very early stage in the furnaces and ovens of the ceramics industry. They found special favour for the construction of the platforms of the wagons of tunnel furnaces. Refractory concrete was found to be definitely superior to normal refractory materials for this purpose [257], [258], [259]. Subsequently high-temperature concretes have been introduced with considerable success for the same purpose in many countries throughout the world. In the last few years refractory concrete has also come to be used for other, stationary, components of the furnaces in this industry.

The duty required of the linings for the wagons of tunnel furnaces is particularly severe, on account of the extremes of temperature cycling and other conditions to which they are subjected [260]. Refractory concrete not only possesses better properties in this respect than normal brickwork, such as superior resistance to abrasion resulting in longer working life; it is also easier and quicker to repair and replace. In addition, if portland cement or blast-furnace cement are used, the cost is less than that of brick linings.

The platforms of furnace wagons are frequently constructed of cast-in-situ concrete, but precast components may also be used for this purpose. It is, of course, essential to choose the correct concrete for the anticipated operating temperatures, to construct it with a sufficient thickness and to allow suitable expansion joints. There is an abundance of technical specifications, design recommendations, and practical experience for this type of work [96], [97], [98], [111], [141], [227], [259], [260], [261], [262], [263]. Two examples of the construction details of the lining for a furnace wagon are shown in Fig. 69.

High-temperature concrete may be used for the wagon linings of any type of tunnel furnace. A large number of examples of this application is available from the brick-making, sanitary ware and porcelain industries. If, however,

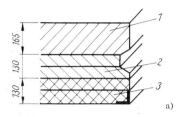 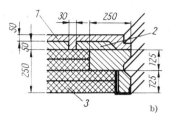

Fig. 69. Construction details for lining of tunnel furnace wagon

a) for 1050 °C; b) with a smaller thickness of the concrete layer, for 1060 °C prefired at 1400 °C (Wilde)
1 Concrete; 2 Lightweight refractory brick; 3 Insulating brick

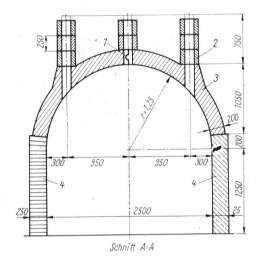

Schnitt A-A

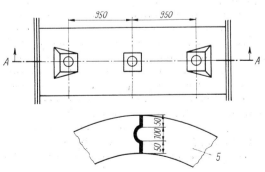

Fig. 70. The chamber of an annular furnace, constructed of aluminous cement (Schmeisser)

1 Expansion joint; 2 Precast units of smelted aluminous cement concrete; 3 Concrete, smelted aluminous cement; 4 Refractory brick or smelted aluminous cement concrete; 5 Hinge at apex between arch-units

the operating temperature of the furnace is relatively low, the necessary ceramic binding of the concrete may be ensured by subjecting the lining to an initial, higher temperature when it is first fired. The use of refractory concrete in this type of installation results in an appreciable increase in the number of cycles which the wagons are capable of withstanding. An example may be quoted from *K. D. Nekrasov* and *G. D. Salmanov* [224], who report an increase in the number of wagon journeys into a furnace from an initial figure of 6 to 8, to more than 100.

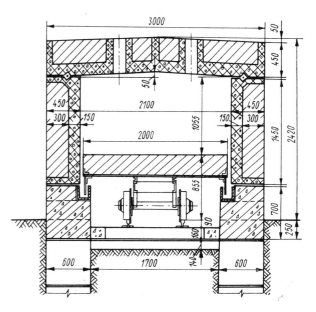

Fig. 70a. Layout of a tunnel furnace for operation at middle-range temperatures. constructed of precast components of refractory concrete [237]

In ceramics furnaces refractory concretes are used both for repairs and for new constructions. In-situ concrete, precast concrete, or a combination of the two are common. The dimensions of the furnace do not have to be altered for this purpose.

Refractory concrete may be used for practically every type of furnace lining, provided the temperatures permit it. Essential components may be manufactured in this way for tunnel furnaces, annular kilns and zigzag kilns. Examples which may be quoted here are a ring furnace roof in West Germany [268], shown in Fig. 70, and a tunnel furnace from Soviet sources [237], shown in Fig. 70a. It is stated in connection with the construction of the latter structure, that only 7 workmen were required for a total of 10 days, whereas if conventional refractory brick linings had been used, 20 men would have been needed for 30 days. The saving in construction time, the increase in productivity achieved and the reduction in cost (in this case 20 to 30%) are clearly evident from these figures. The binding agents employed included both hydraulic binders, such as aluminous, portland and blast furnace cements, and waterglass [264]. The aggregate chosen was suitable for the operating temperatures. The following references describe some of the examples detailed in technical literature on the subject: burner kerbs and furnace walls (constructed partly and completely of refractory concrete) [190], [265]; furnace roofs [137]; feeding openings, furnace soles, extract ducts, doors, and so on [63], [220], [229], [266], [267], [268], [269].

6.5. The Glass-making Industry

Very little is recorded about the use of fire-resistant concrete for furnace linings in the glass-making industry, although there would appear to be many possible applications of the material in this branch of technology. It should be possible, for example, to use concrete for the linings of gas extract ducts or of parts of regenerative furnaces, in the same way as it is utilised for these purposes in other industries. The first trials in glassworks, such as the Special Glass Unit at Weisswasser and Plate Glass Factory at Torgau, in which linings were constructed of refractory concrete for duct walls in the sub-furnace, for various blocks and for chimney ducts and similar connecting ducts in tank furnaces and tempering furnaces, have so far presented no problems.

Studies are being conducted in the U.S.S.R. of problems associated with the use of high-temperature resistant concretes in the glass industry. Experimental furnaces for glass-smelting constructed in this way are already under test [270]. It is anticipated that it will be found possible to construct roofing sections of glass tanks out of fire-resistant concrete, provided that suitable alkali-resistant concrete mixes can be developed [227].

An experimental glass-making furnace erected in Hungary, in which the roof was constructed of special fire-resistant concrete made with corundum and a phosphate binding agent, is reported to have proved satisfactory in service [213], [214].

The smelting furnaces used for the manufacture of mineral wool can be considered as similar in many respects to those for glass-making. Something will therefore be said on this subject here. *Ja. P. Mostovoj* and his colleagues [298] have described the successful use of refractory concrete in a slag-smelting furnace. The original lining of this furnace, constructed of standard chrome ore-magnesia bricks, had suffered from excessive corrosion of the joints in the brickwork, on account of the continuous variation in the level of the molten material. The life of this type of lining was consequently short. Chrome ore-magnesia concretes, made with either portland cement or periclase cement, were found to be quite satisfactory in service, in spite of the high operating temperatures of 1400 to 1450 °C. These concretes had a t_a-value of 1300 to 1350 °C and a t_e-value of 1350 to 1530 °C and the cement:aggregate ratio was 1:5 for the portland cement mix and between 1:1 and 1:2 for the periclase mix. The resistance to attack by slag is good, as the chemical reactions lead to the formation of a crust, which inhibits further slag attack and mechanical abrasion. The re-use of broken magnesia and chrome ore-magnesia bricks, together with the method of construction employed, resulted in a 50% reduction in the costs of construction and of repair of the furnace.

14*

6.6.   The Cement and Lime Industry

The use on a large scale of fire-resistant concrete really dates back to the use in the cement industry of clinker concrete linings, which were discovered by *F. Valeur* [271] and patented by him as linings for rotating cement kilns. This clinker concrete is nothing more than a granular, graded clinker, which is bonded to a cast-in-situ structure or to cast bricks by means of cement. The material has been used in this way to a considerable extent in cement works and may therefore be regarded as the forerunner of present-day refractory concrete. Clinker concrete will not be discussed in more detail here, but interesting information on the subject may be found in the publication by *H. Kühl* [16; pages. 487, 535].

Clinker concrete was replaced some decades ago by high quality refractory brick; the use of the latter material has persisted until the present day, but developments are once again leading to the use of concrete, now that concrete mixes are being developed which possess the necessary thermal and mechanical properties for resisting the severe conditions found in rotating kilns. The technical literature published on the subject indicates that encouraging results are being obtained in this respect, although the subject is still in the experimental stage.

The possibility of using linings of refractory concrete in rotating kilns had already been tested in a number of cement works in the U.S.S.R. as early as 1939, without, however, arriving at conclusive results. The concretes used were of aluminous cement and chrome ore. After the second world war the subject was opened up again; refractory concrete was used in ancillary plants to the main furnace, for instance in heat exchangers [272], [273], and the question of employing this material for the lining of kilns was also raised again. *A. K. Beljaev* and his colleagues, for example [198] ,[274], described the conditions that must be satisfied if fire-resistant concrete was to be used successsfully in cement kilns. Concrete then appeared to be especially suitable for lining stationary components of the plant and those sections in which the most severe conditions were not found, such as drying and calcining zones. The use of concrete was considered only to a limited extent for the sintering zone, since the operating conditions here are exceptionally severe. Magnesia concretes bound with waterglass might, for instance, be suitable for the latter purpose. After the good results recently obtained in high-temperature metallurgical furnaces using magnesia-bound concretes, such as sintered magnesia, chrome ore and titanium alumina slag with Sorel's cement, experiments have now been carried out in the U.S.S.R. with the lining of the sintering zones of rotating cement kilns as well, with this type of material [209].

Information exists in the published technical literature about the lining of shaft lime kilns, in particular the preheating and discharge zones, with high-temperature resistant chamotte concretes [275], [276]. The resulting advantages include improved strength and increased durability as well as economy and an improvement in the output of the furnace. Concrete linings have also proved to be satisfactory in lime kilns subject to cycles of temperature, where the refractory material is maintained for some time at a temperature of 900 to 1100 °C and is then cooled down. Slag — portland cement — chamotte concrete has been used for this purpose in the Kuznesk Works, withstanding more than three years under operating conditions [315]. The damage which occurred resulted essentially from the penetration of limestone into the joints and the consequent forcing of them apart under the effects of heat; the blocks themselves were hardly affected.

Similar satisfactory results have been obtained in the lime industry in the German Democratic Republic. The working life of the linings of the preheating zones of shaft lime kilns can be increased several times over by the substitution of prefabricated concrete units or bricks for the normal chamotte lining. The concrete in this case was composed of portland cement, chamotte 0 to 2 mm (0,08 in) and chamotte 2 to 20 mm (0,8 in) in the proportions 10:12:10, one of the principal reasons for its superior performance being its high wear-resistance. The utilisation of broken chamotte brick and the small incidence of repairs required result in the cost of these linings being only about 50% of that of normal chamotte brickwork.

6.7. Coke Ovens and Gasworks

The use of fire-resistant concretes is fairly widespread in the coke industry. Refractory concretes are used here for the manufacture of a great variety of furnace components. It is reported by *T. D. Robson* [277] and also by *K. D. Nekrasov* and *G. D. Salmanov* in their survey of the subject [214] that it is common in England for foundations, doors, keystones, door-frames, charging openings and viewing windows of coke ovens to be made of refractory concrete, as well as the wagons used for the cooling of the coke. It is also common in gasworks to use heat- or fire-resistant concretes for lining retorts, pipes, cooler floors and reduction chambers and for the construction of foundations.

As can be seen from recent papers, the use of refractory concretes is being continuously extended in the field of coke ovens in the Soviet Union. Doors had already been constructed of refractory concrete in 1938 [278]; this practice is normal today [219], [220]. The upper closure units of large-sized coke ovens have been constructed of refractory concrete for a number of years in the Kuznesk Works [315]. The mix comprises 310 kg of portland cement (quality 400) and 1490 kg of chamotte (of less than 12 mm (0,48 in) grain size) per cubic metre (522 and 2505 lb/yd³) of concrete. The very fine fraction, of less than 0,09 mm (0,036 in) grain size, is equal to 37 to 52% by weight of the cement. The reinforcement is of steel mesh. The working life of one of these units is from 6 to 7 months. In this situation waterglass concrete did not prove to be so satisfactory as portland cement concrete. Heat-resistant reinforced concrete has also recently been adopted for the construction of the foundations of batteries of coke ovens [279], [280].

J. Hodgson [281] has published a summary describing the use of fire-resistant concretes in Western countries. Refractory concrete has also recently been introduced into the construction of coke plants in the People's Republic of China [256].

According to *H. Mitusch* [98] fire-resistant concrete has proved to be exceptionally satisfactory for furnace doors in coke plants, since concrete possesses a much longer working life than normal brick-lined doors, on account of its high stability under fluctuating temperature. This author also recommends refractory concrete for lining rising pipes and headers.

Precast units of refractory concrete have also been used with success in coke ovens in the German Democratic Republic [309]. The dividing walls in the substructure of regenerative furnaces are being constructed of this material, as a replacement for quartz chamotte, and also cut-off gates. Concrete has also been used experimentally in recuperators and in the flues of degasification chambers. There is a tendency in certain temperature regions, not yet fully explained, for CO to penetrate and carbon to be precipitated; this may lead to deterioration of the concrete.

For gasification at high temperatures, above 1300 °C, the use of special refractory concretes which possess high values of thermal conductivity and good fire-resistant properties appears to offer promising possibilities, as far as can be judged from the first test results [297].

6.8.   The Power Industry

The introduction of refractory concretes into the electrical generating industry has taken place fairly rapidly and without any exceptional difficulty arising. The most important applications so far have been to the construction of power station plant in general and to boilers in particular. A great variety of possible applications presents itself in this field, especially as excessively high temperatures are not encountered.

In the U.S.S.R. it has been common practice for some time to use fire-resistant concrete for the construction of linings in boiler plant [214], [282]. The concretes used are generally made with portland or aluminous cement. The same is also true for the boilers of power stations in Hungary [214]. In the Soviet Union modern, large-scale boilers with a rating of 120 tons/hour of steam and operating pressures of 110 atmospheres (1620 psi) are now provided with linings constructed of a combination of different materials, of which refractory concrete constitutes an appreciable part. Various materials are used, including chrome-ore concretes bound with waterglass, lightweight concretes made with diatomaceous earth, asbestos and either aluminous or portland cement, and mixes for use as renderings which are made from aluminous cement, clay, asbestos, chamotte and waterglass. The working life of these lining materials is found to be good. Detailed information about them is available from *I. Ja. Zalkind* [197].

Codes of Practice have been published in the Soviet Union which cover the use of fire-resistant concretes in boiler plants [283].

In the German Democratic Republic much research has been carried out on concretes for use at high temperatures, directed towards demonstrating the suitability of refractory concrete made with cement as the binding agent for power stations and boilers. Aluminous cement-chamotte concrete was used in the Trattendorf power station for the first time in 1958 for lining the roof of the boilers and the smoke extract shafts [107]. From 1959 onwards, precast units of portland cement chamotte concrete have been used to an increasing extent. Examples include Böhlen Power Station, the Schwarze Pumpe Works, and many others. Among the structures involved were hanging roofs, slabs, gates, hoppers and headers.

The performance of refractory concrete linings has proved to be extremely good [107], [229], [308], so that considerable progress is now being made in utilising them in this industry. In comparison with normal chamotte linings they possess considerable economic advantages, not only by replacing the very large number of different shapes of brick previously in use by a few standardised precast units, but also by effecting an increase in construction efficiency of about 35% in the case of boilers and a reduction in cost of at least 15%. More detailed information may be obtained from the paper by *Haubenreiszer* [308].

A Standard is available in the German Democratic Republic covering precast units of refractory concrete for water-tube boilers [4].

6.9. The Chemical Industry

There is considerable scope for the use of heat-resistant concretes in the chemical industry, since the temperatures encountered in most of the plant are in the middle range. The principal problems therefore relate less to the thermal properties of the concrete and more to its mechanical properties, resistance to wear and chemical resistance, and so on.

Heat-resistant concretes made with waterglass have come to be the preferred material in the Soviet Union for the linings of certain items of chemical plant. Examples are roasting furnaces (calciners) for pyrites and other materials and soda regeneration plants, as described by *K. D. Nekrasov* [41]. A later paper published by the same author jointly with *G. D. Salmanov* [224] reports that up to the time at which it was written the linings of more than 40 furnaces used for the roasting of iron pyrites and the ores of non-ferrous metals had been constructed of waterglass-chamotte concrete and had proved successful in operation. The reduction in cost, in comparison with similar plants in which refractory bricks were used, was of the order of 40 to 50%. The operating life was increased two- or threefold.

Normal refractory concretes made with hydraulic binding agents are, however, also sometimes used. The first experiments in this connection were carried out a long time ago in soot plants. Here the foundations were constructed of portland cement concrete containing pumice aggregate and crushed limestone [27].

In the oil industry in the U.S.A. fire-resistant concretes have now been successfully used for a number of years in refinery boilers [160], [180], [284]. One of the consequences of this application has been a series of systematic investigations which have provided results which are of interest to the whole of the refractory concrete industry. These results have already been discussed when dealing with the properties of concretes. Refractory concretes made with portland cement and chamotte have also been used in the Soviet Union in oil-refining plants [190], [227], [235], [285]. In one oil distillation plant one of the structures so constructed was the suspended lining of the tubular preheating unit which was constructed of fire-resistant concrete. The cost of the lining of this tubular furnace was approximately 55% lower than that of a normal lining of chamotte brick. Similarly, instead of the expensive preformed refractory brick normally used, the fire bars of a low-temperature carbonising furnace were constructed of blocks of heat-resistant concrete having dimensions of $1 \times 0,6 \times 0,45$ m ($3^1/_4$ ft $\times 2$ ft $\times 1^1/_2$ ft) [224].

A paper by *M. A. Vejcman* and *I. M. Ceperskij* [286] provides information about the experimental construction of a flameless furnace in the oil refining industry from refractory concrete.

Particular mention should be made of the practice in the Soviet petrochemical industry of using fire-resistant sprayed concrete for lining connecting pipelines and other plant items as a protection against the effects of heat, corrosion and erosion. This type of plant handles the heavy oil residues of a catalytic pyrolysis process and of continuous carbonisation, temperatures of up to 1000 °C being encountered. Some form of reliable thermal insulator is essential. In other types of reactors and regenerators the lining materials may be subjected to temperatures of 500 to 600 °C, pressures of up to 50 atmospheres (about 700 psi) and the effects of aggressive chemicals. The results of several years of operating experience have demonstrated that a lining of fire-resistant mortar and monolithic concrete proves very serviceable under these conditions.

The mix used for the protective concrete lining will depend upon the function which it has to fulfil. Examples include normal refractory concretes of average density and strength, made with aluminous cement, portland cement or pozzolanic portland cement and diabase, lightweight chamotte, diatomite or chamotte as the aggregate. Other types include insulating concretes of average density, possessing satisfactory mechanical strength as well as the necessary insulating properties. Other types again are lightweight concretes of low thermal conductivity and concretes possessing a high degree of resistance to wear and erosion. The linings are composed of at least two layers, sometimes more, of sprayed concrete. A common practice is to form the first layer of 25 to 35 mm (1 to 1,4 in) of a mix of cement, diabase powder and ground chamotte sprayed on to a coarse, thick wire reinforcing mesh. This is known as the anti-corrosion layer and is followed by an insulating layer 125 to 175 mm (5 to 7 in) thick. For the third layer, a wear-resistant mix is sprayed on to a further mesh, forming a protective surface. The whole is then hydrothermally hardened by means of steam fed into the plant at a temperature of 160 °C and a pressure of 5,4 atmospheres (77 psi).

G. Franke and *F. Kanthak* [107] have described the use of concretes made with smelted aluminous cement and portland cement for lining the furnaces of coke plants in the VEB tar-producing plant at Rositz. The experience obtained with this form of lining is encouraging and justifies its further use.

Heat-resistant concretes are also suitable for lining pressurised items of plant which operate at elevated temperatures. It is reported that an autoclave has been constructed in the Peoples' Republic of China, with dimensions of 3 metres (10 ft) in length $\times 1,5$ m (5 ft) in diameter. This comprises prestressed reinforced concrete rings and operates at pressures up to 12 atmospheres (170 psi).

Refractory concretes have been successfully employed for the graphiting furnaces of the electrode factory at Novočerkassk. More detailed information can be obtained from a brief paper by *A. E. Issers* [227]. Precast concrete units have also been used in resistance furnaces in the nitrogen fertilizer factory at Piesteritz. Here blocks of portland cement concrete, each having weights of up to 3,3 tons, were used for the end walls. These blocks have a working life of at least one year, compared with the three months life which was usually obtained with the normal lining materials previously used [325].

As early as the end of the 1940s refractory concrete components made with aluminous cement were being used in the U.S.A. in reaction furnaces in which phosphorus was extracted from phosphates. The durability of this type of lining was found to be satisfactory. More recently, however, a change has been made to waterglass concrete for this purpose [220].

6.10.   Airports and the Aircraft Industry

Mention has already been made, in the discussion on foundries, of the fact that heat-resistant concretes are suitable not only for use in furnaces and similar plants which operate at elevated temperatures, but that they may also be used in place of normal (dense) concretes for the construction of slabs and similar structures which, because of their special operating conditions, are subject to sudden temperature variations. This is becoming increasingly important in relation to the take-off areas of jet aircraft, where localised areas of the concrete are subjected to the effects of the exhaust gases of the jets. The problem of the heat-resistance of such areas will become increasingly acute, as the technology of vertical take-off becomes more fully developed. This is also true for rocket launching pads.

Questions of this type have already been under consideration for a number of years, as can be seen from a study of the technical literature. One example is the investigations by *P. H. Petersen* [287] into the damage caused by jet aircraft to landing grounds. In a detailed American review of the subject by *H. G. Protze* [288] the view is expressed that take-off areas subjected to high temperatures should preferably be constructed of heat-resistant concrete made with aluminous cement. In a similar manner, the serviceability of aluminous cement concrete has been proven in test cells for jet engines [320]. In test facilities of this type the surfaces of the walls are raised to a temperature of about 650 °C in 8 seconds by the action of the hot gases; up to 5000 tests of this type may be carried out in one year. Although this lies in the critical temperature range for cement-bound concretes, the use of aluminous cement concrete is recommended; since the heat is applied to one side only, it penetrates to only a limited depth. In making this class of concrete, it has been found helpful to use additives to control the setting time, and to introduce air--entraining agents to improve the workability.

6.11.   Nuclear Engineering

In nuclear reactors, radiation is present in addition to large quantities of heat. Special fire-resistant concretes are therefore sometimes recommended, designed to serve a dual purpose. In this connection *P. P. Budnikov* and *V. G. Savel'ev* [65] suggest the use of barium monoaluminate concrete with chamotte or chrome ore-magnesia as the aggregate. Attention has been drawn in an anonymous article [289] to an aluminous cement concrete made with the use of a newly-developed aggregate, described as "Alag", which is said to possess excellent radiation shielding properties. Nothing more is said about the nature of this material, apart from an account of some of the properties of the concrete made from it.

6.12.   Flat Construction

Heat-resistant concrete is useful not only for industrial purposes, but also for the construction of domestic furnaces. *Nekrasov*, for example, describes a standard domestic stove, assembled from precast units of portland cement concrete [6]. A brochure by *L. S. Bordzenko* [324] describes and explains in detail the design of domestic stoves made of precast concrete blocks and of domestic chimneys consisting of one or two separate chimney flues. This booklet also contains practical instructions about manufacturing procedures and other technical information relating to these articles. The mechanisation possible with this type of prefabricated work allows considerable economies to be made in the cost of materials, in construction time and in the overall price. These stoves and chimneys have proved satisfactory in use.

In the German Democratic Republic, chimney elements have now been used in considerable numbers in the industrialised construction of flats. The material used is heat-resistant concrete made with broken brick and portland cement. More recently, it has been found that equally satisfactory and sometimes better results can be achieved if crude copper slag is substituted for the broken brick [296]. This modification results in appreciable economies. A typical concrete mix has already been discussed in Section 3.4.3.6.

Details are also available in brochures published in other countries by the makers of refractory concretes, relating to designs of fire-places and complete furnace linings. The Société Anonyme des Chaux & Ciments de Lafarge & du Teil may be quoted as an example.

7. Some Economic Aspects of the Use of Refractory Concretes[1]

The overall economic advantages of using refractory concrete are undisputed. This is becoming increasingly clear from every paper published on the subject. The quantity of information in technical literature relating to economics is so large and often so detailed that separate cases cannot be quoted here. It is preferable therefore to deal only with some of the fundamental aspects

It is stated by *K. D. Nekrasov* [6] and confirmed by many other authors that the reduction in the construction costs of refractory plants may be anything from 10 to 50%, depending upon the type and form of the refractory products employed. A large number of examples were provided at the Conference on Research, Design, Construction and Operation of Refractory Plants using heat-resistant concrete and reinforced concrete held in Moscow in 1960 [291], which illustrated conclusively the superiority in economic terms of heat-resistant and fire-resistant concretes over normal fired materials, in specific cases. It was stated in a Report of the Communist Party of the Soviet Union in November 1962, in connection with the development of the Soviet refractory industry, that the use of heat-resistant concrete in the construction of refractory plants had reached a figure of 400 000 cubic metres (521 600 cubic yards) or 20% of the total of all refractory building materials used in the Soviet Union and had resulted in a saving of 25 million roubles.

According to a Polish publication [123] the use of heat-resistant concretes has effected a reduction on capital cost of between 50 and 70%, compared with the cost of normal refractory materials.

In the German Democratic Republic, figures supplied by the industries using refractory materials indicate that savings totalling more than 1 million marks per year may be possible as a result of the use of precast components of refractory concrete alone, in the metallurgical, chemical, electrical, glass and ceramic industries. In addition, in some cases, there will be higher availibility of the existing plant, as a result of the reduction in the time required for effecting repairs and installing linings to furnaces and of the increase in durability.

In this chapter a discussion of a number of special considerations relating to in-situ and precast construction will be used to illustrate the possible applications of both types of construction from the point of view of industrial economics.

---

[1] The material used in writing this chapter includes information and results obtained from a conference held at Freital, near Dresden, in November 1963.

7.1.   In-situ Concrete

The use of cast-in-situ concrete permits the mixing and handling of the concrete to be carried out directly adjacent to the work. Alternatively, dry mixes may be used which have been prepared by a specialist manufacturer and delivered in this condition to the site. If the required concrete mix can be obtained from the range available from the specialist works, the latter procedure is to be preferred. Preliminary trials and tests to establish the properties of the mix are then for the most part unnecessary.

The use of monolithic concrete for the construction of refractory linings enables both new linings and repair works to be constructed to a substantial thickness. The amount of work entailed is comparable to that required for normal concreting and is thus considerably less than that needed when refractory brick linings are employed. Statistics from Poland indicate that the capitel costs incurred when heat-resistant concrete mixes are used are from 30 to 50% lower than those for chamotte linings [103]. An additional advantage is the facility with which almost any desired structural form can be constructed. If rapid-hardening cements are used the speed of construction will be high and rapid re-use of the shuttering will also be possible. To quote an example, a ring furnace with a firing channel 69 m long and 2,4 m wide (approx 228×8 ft) was stripped and relined with heat-resistant concrete by 15 inexperienced workers in 50 days [141].

The casting of concrete linings in-situ for the construction of complete plants will, however, tend to be replaced in the future to an increasing extent by the use of precast units. It may be expected that cast-in-situ concrete will continue to find favour, as heretofore, for repair work. On the other hand, in-situ concreting enables the number of different shapes of components to be reduced, a point of particular significance when only a small number off is required [229]. The furnace designer should always endeavour to design precast components in such a way that their fabrication can be mechanised, though the requirements of the plant will usually limit the extent to which this can be achieved.

7.2.   Precast Components

The utilisation of precast components constitutes the most rational system of concrete construction so far devised which can be adopted for high-temperature concretes. The introduction of mechanised construction to conventional civil engineering and building, together with the use of standardised concrete elements of large dimensions, made a steep rise in productivity possible; similar methods of prefabrication should therefore produce favourable results in the construction of refractory linings. The use of prefabrication will always result in faster construction times than in-situ construction, while enabling a high standard of quality to be achieved, though in-situ construction will remain suitable for many purposes; prefabrication is justified whenever a large number of components of standardised shape are required [229]. Prefabrication is of use not only for the initial construction of refractory plants, but also, equally, for rapid and organised repairs to plant which is subject to heavy wear. It enables down-time and loss of output to be reduced to a minimum.

The normal output per man-shift for the manufacture of chamotte bricks by the wet process may be taken as 0,7 tons. The current figure for the fabrication of precast units of refractory concrete in the VEB Silica and Chamotte Works at Rietschen is 2,5 tons per man-shift. This increase in productivity by a factor of 3,5 means that, whereas 140 men are required for a daily production of 100 tons of chamotte bricks, only 40 are needed for a similar quantity of precast units (quarrying, raw chamotte production and workshops are excluded from this comparison). A figure of 4,0 tons per man-shift appears to be possible when future improvements to the production-line have been carried out.

It is desirable for the manufacturer to produce only a certain number of precast units of standard form. The weights of the units should also be suited to the requirements of both the manufacturer and the user. The Rietschen Works, for example, therefore require that a component weight of more than 100 kg (220 lb) should involve a considerable number of units of the same type. Units up to weights of 20 kg (44 lb) are manufactured in a great variety of shapes and sizes. The maximum component weight produced at present is 2 tons, but it will be possible to increase this in the future to 5 tons. Components of comparable weights are also manufactured in the Soviet Union. At present the limits in this direction are probably set by the means of transport available and the storage at the user's works, at least so far as the existing plant and buildings are concerned. If large-sized components are to be utilised, and mechanised methods of erection adopted, this will, of course, mean that the user will require new plant for transporting and lifting the prefabricated units. The investment in this temporary plant will be amortised in the shortest time when it is used in conjunction with refractory plants which are subjected to heavy and rapid wear. Spaces will need to be left in certain locations for the handling equipment; these will be filled in later with cement mortar.

When large numbers of components of the same type are required, it may appear to be economical to manufacture them on the site of the refractory plant; this will, however, be only a temporary solution. A specialist manufacturing plant is always able to employ up-to-date methods and techniques and is in a position to improve the existing technology. When it is considered that in the future certain prefabricated units may be formed under high pressure and that methods of rapid-hardening and consequently of improving the quality of components, even if they are of high-temperature concrete, will be more common, it will be appreciated that the expense involved can be justified only in the case of a specialised manufacturing works. As in other branches of industry, the mechanisation and automation of production processes can only bring high returns when the number of standardised components to be manufactured is large.

The use of refractory concrete always results in a reduction in the number of joints required, regardless of the method of construction employed. It is well known that the jointing mortar is the weakest part of a refractory structure. In addition, the number of skilled refractory bricklayers is reduced to a minimum, since relatively few skilled workers are required for the supervision and direction of the less qualified trades employed in the mixing and placing of concrete or in the fixing of precast components.

7.3.　　Considerations of Price

The competitiveness of high temperature concretes is not only influenced by the points just discussed; other factors which affect the economics include the way in which the concrete is utilised and the purpose for which it is intended, its operating life and, finally, its composition; that is to say, the nature of its constituent materials. It will therefore be useful to consider some of the price comparisons in the light of these variables.

The type of binding agent used is the major factor influencing the cost of the commoner concrete mixes made with chamotte, such as raw (coarse) chamotte, types I and II in accordance with TGL 99-30 [5]. Normal commercial aluminous cement is approximately 10 times as expensive as portland cement; for standard concrete mixes, therefore, the unit cost of concrete made with aluminous cement will be about twice that of concrete made from portland cement, or about one-and-a-half times in the case of precast components. This comparison is based upon the present price lists in force in the German Democratic Republic [292]. When special cements are used (such as smelted aluminous cement), the differences are somewhat larger.

In these circumstances it is therefore essential to establish which type of binding agent is justified for a particular operating temperature. It would, for example, be uneconomic to use aluminous cement for concrete required to withstand temperatures no higher than 1150 °C, since portland cement or perhaps even slag cement would be satisfactory for this purpose and would be far cheaper. Similarly, when an engineer is designing his own mixes, he should take into account the effect of the cement content upon the cost of the concrete. Since the strength at high temperatures is improved very little by high cement contents, the minimum possible cement content can be used. The effect of this point upon the price should not be underestimated; for example, if the cement content of a mix containing 20% of aluminous cement is increased to 30%, the overall price is increased by 32%.

The influence of the binding agent upon the cost of a fire-resistant concrete becomes less significant when refractory materials of high quality are introduced into the aggregate fractions. The costs of sillimanite and corundum, for instance, are respectively about 10 and 15 times that of chamotte. They are thus approximately twice as high as the cost of industrial aluminous cement.

The material used as the aggregate can therefore also have a decisive influence upon the overall cost. In average mixes, particularly those cast in-situ, an economic advantage can be obtained by the use of broken chamotte or old broken material (from stripped-down plant). It is, of course, self-evident that the aggregates, like the cement, must be chosen to suit the required operating temperatures.

For high-temperature purposes only high-quality materials should, of course, be used. No special advantage is obtained, however, by utilising these materials in the middle temperature range — below 1000 °C. In the latter case, much cheaper aggregates, such as broken brick, blast furnace pumice, slag and basalt, answer the purpose quite satisfactorily. The cost of such materials will, moreover, already have been borne by the industry of which they are by-products, leading to further reductions in the cost [136].

If the prices of refractory concrete mixes and precast units based on chamotte aggregates are compared with those of equivalent refractory materials (especially chamotte), the following relationships may be deduced: for normal rammed or trowelled mixes concrete is clearly more expensive; it does, however, possess the advantage of earlier strength in the unfired state. When a comparison is made with unfired chamotte bricks of similar quality, the position of concrete, considered either as a mix or in precast form, will vary depending upon the type and quantity of cement used and the size of the work to be constructed. A final comparison can be made only when the details of the required application are known. Compared with fired, general-purpose, chamotte bricks (TGL 4323) refractory concretes are more economical, especially in those cases where in-situ concreting is possible or where precast units of large size (from 40 to 500 kg (88 to 1100 lb)) can be utilised in place of normal or small bricks. Thus 1 ton of normal or arch bricks of quality III and IV costs about the same as 1 ton of concrete containing 20 or 25% respectively of aluminous cement and approximately twice as much as the corresponding portland cement mix, or approximately the same as precast, aluminous cement units exceeding 500 kg ($^1/_2$ ton) weight or precast, portland cement units up to 40 kg (88 lb) weight. Mechanically, nonplastic pressed, profiled bricks of qualities III and IV of weights greater than 40 kg (88 lb) are more expensive than concrete mixes made with aluminous cement and are approximately the same as large precast aluminous cement units; here it should be noted that the unit price of precast units of 500 kg ($^1/_2$ ton) weight is about 26% lower than that of units of 40 kg (88 lb) weight or less. The relationship for portland cement concrete is similar. Plastically formed profiled bricks cost about the same as concrete mixes containing 25% aluminous cement (twice that of portland cement mixes) and are of the same order as precast units of aluminous cement greater than 500 kg ($^1/_2$ ton) or portland cement units up to 40 kg (88 lb).

Even though a direct comparison between refractory concrete and conventional refractory bricks is not always possible, in those cases where the prices are of the same order the other advantages of high temperature concrete favour the competitiveness of cement-bound concretes in furnaces and refractory plants. In separate cases the prices of refractory concretes are lower than those of the refractory materials hitherto in general use, but even in those cases where these advantages do not obtain, the additional expense is counterbalanced by many other advantages, as has been explained in detail alsewhere.

7.4. Summary of Economic Considerations

In this final chapter it has, of course, only been possible to consider a number of the essential economic aspects relating to the choice and use of refractory concrete as an alternative to equivalent refractory bricks.

There now follows a summary of the most important factors which give to high temperature concretes the superiority and the advantages which characterise them in comparison with traditional refractory materials:

(a) The use of concrete in cast-in-situ work permits the construction of a monolithic lining and avoids the need for complicated profiled bricks; in addition, the simplification of the work results in a saving in construction time and costs. Compared with normal rammed and trowelled mixes, concretes possess high strength, even in the unfired state.

(b) Precasting makes possible the introduction of industrialised methods of construction to the building of furnaces, by the use of large-sized components.

(c) The manufacture of precast units leads to a considerable increase in productive efficiency and facilitates the mechanisation of production procedures in the manufacturer's works. A saving in production costs also arises as a result of the elimination of the procedures of pressing, drying and firing and from a reduction in the quantity of plant and equipment required.

(d) On the part of the customer or purchaser there is a saving in the size of labour force needed and a reduction in the time required for the construction and repair of refractory linings. A saving in cost and an improvement in productivity results from this. The use of refractory concrete therefore has specially marked economic advantages for units of plant which are subjected to heavy wear.

(e) In comparison with refractory bricks, refractory concretes often possess higher resistance to chemical attack, abrasion and temperature cycling; the durability of linings can be extended considerably as a result.

References

In this List of References the titles have been given in their original form. Where an English translation is given in brackets after a title in a foreign language, it should not be assumed that an English translation of the work necessarily exists.

[1] *Konopicky, K.:* Feuerfeste Baustoffe. (Refractory building materials.) Düsseldorf: Verlag Stahleisen m.b.H. 1957.
[2] *Harders, F.,* and *Kienow, S.:* Feuerfestkunde. (Refractory News). Berlin-Göttingen-Heidelberg: Springer 1960.
[3] *Rabe, G.:* Silikattechn. (Silicate Technology.) **11** (1960) No. 9 (Appendix); Worksheet No. 152.
[4] TGL[1] 9356. Feuerbeton-Fertigteile für Wasserrohrkessel. (Precast refractory concrete units for water-tube boilers): Sept. 1961.
[5] TGL 99-30. Feuerbeton-Gemenge für pyrotechnische Anlagen. Technische Lieferbedingungen (Entwurf). (Mixes for heat-resistant concretes for use in refractory plants — Design Specification.) July 1963.
[6] *Nekrasov, K. D.:* Šaroupornyj beton. (Heat-resistant concrete.) Moscow: Promstrojizdat 1957.
[7] Spravočnik po ogneupornoj kladke promyšlennych pečej. (Handbook of refractory linings for industrial furnaces.) Moscow: Gosstrojizdat 1960.
[8] Instrukcija po technologii prigotovlenija i primenenija žarostojkich betonov. (Direct ive on the manufacture and use of heat-resistant concretes.) pp. 156—161. Moscow: Gosstrojizdat 1961.
[9] Techničeskie uslovija na tonkomolotye dobavki i zapolniteli dlja žarostojkich betonov. (Specification for finely ground aggregates and additives for heat-resistant concretes.) MRTU 7-3-60. Moscow: Gosstrojizdat 1961.
[10] *Budnikov, P. P.,* et al.: Technologie der keramischen Erzeugnisse einschließlich der feuerfesten Baustoffe. (The technology of ceramic products, including refractory building materials.) Berlin: VEB Verlag Technik 1953.
[11] *Zapp, F.,* and *Dramont, F.:* Ber. dtsch. keram. Ges. (Report of the German Ceramics Society.) **32** (1955) No. 4, p. 97.
[12] *Mitusch, H.:* Die Abhängigkeit der Eigenschaften feuerfesten Betons von den Beziehungen zwischen Zuschlagstoff und Tonerde-Zement. (The effect upon the properties of refractory concrete of the interaction between the aggregate and aluminous cement.) Dissertation Clausthal 1958.
[13] *Braniski, A.:* Tonind.-Ztg. (Clay Industry News.) **85** (1961) No. 6, p. 129.
[14] *Schulze, W.:* Der Baustoff Beton und seine Technologie. (Concrete as a building material and its technology.) 2nd Ed., Berlin: VEB Verlag für Bauwesen 1961.
[15] *Reinsdorf, S.:* Leichtbeton. (Lightweight Concrete.) Vol. 1. Berlin: VEB Verlag für Bauwesen 1961.

---

[1]) TGL: Standard of the German Democratic Republic.

[16] *Kühl, H.:* Zementchemie. (The Chemistry of Cement.) Vol. 11, 3rd Ed. Berlin: VEB Verlag für Bauwesen 1958.

[17] *Barta, H.:* Stavivo **33** (1955) No. 6, p. 193.

[18] *Stephan, E.:* Tonind.-Ztg. (Clay Industry News.) **78** (1954) No. 1/2, p. 11.

[19] *Platzmann, C.:* DRP 438264 (1925).

[20] *Kestner, P.:* Engl. Pat. 276185 (1926); USA-Pat. 1573072 (1926).

[21] ...: Das Betonwerk (Concrete). **19** (1931) p. 590.

[22] *Coss, H. T.,* and *N. J. Cent:* Ceram. Age 1932, No. 6, p. 212.

[23] *Custodis, A.:* DRP 663342 (1933).

[24] *Braniski, A.:* Tonind.-Ztg. (Clay Industry News.) **59** (1935) p. 143.

[25] *Czernin, W.:* Ber. dtsch. keram. Ges. (Report of the German Ceramics Society) **15** (1934) p. 463.

[26] *Kind, V. A.,* and *S. D. Okorokov:* Stroitel'nye materialy, Moscow; Gosstrojizdat 1934, quoted in [6].

[27] *Aleksandrov, I. A.:* Stroit. prom. 1936, No. 14; quoted in [6].

[28] *Glebov, S. V.,* and *E. A. German:* Ogneupory 1937, No. 9; 1938, No. 7; quoted in [6].

[29] *Gurvic, I. E.:* Stroit. prom. 1937, No. 7; Cement 1938, No. 12, quoted in [6].

[30] *Movsevic, I. L.:* Ogneupory 1938, No. 8; quoted in [6].

[31] *Rojzen, A. I.:* Novosti techniki 1937, No. 18; quoted in [6].

[32] *Ruščuk, G. M.:* Puccolanovye cementy. VNIC, Leningrad 1936; quoted in [6].

[33] *Budnikov, P. P.,* and *D. Z. Il'in:* Cement 1937, No. 7; quoted in [6].

[34] *Hussey, A. V.:* Ind. Engg. Chem. **56** (1937) p. 53.

[35] *Lepingle, H.:* Verres et Silicates **8** (1937) p. 403.

[36] *Möser, A.:* Tonind.-Ztg. (Clay Industry News) **63** (1939) p. 762.

[37] *Giles,, R. T.:* Metals and Alloys **5** (1934) p. 28.

[38] *Giles, R. T.:* Am. Ceram. Soc. Bull. **18** (1939) No. 8, p. 326.

[39] *Nekrasov, K. D.:* Žarostojkie betony kak zameniteli ogneuporov. (Heat-resistant concretes as refractory building materials.) Moscow: Stojizdat 1943.

[40] *Nekrasov, K. D.:* Ogneupornye betony, ich svojstva i primenenie. (Refractory concretes, their properties and use.) Moscow: Strojizdat 1949.

[41] *Nekrasov, K. D.:* Opyt primenenija žaroupornogo betona v teplovych agregatach. (The use of heat-resistant concrete in refractory plants.) Moscow: Gosstrojizdat 1957. (Scientific publications of the Academy for Building and Architecture of the U.S.S.R., Vol. 1.)

[42] *Nekrasov, K. D.,* and *A. P. Tarasova:* Žaroupornyj chimičeski stojkij beton na židkom stekle. (Heat and chemically-resistant concrete made with waterglass.) Moscow: Goschimizdat 1959. (from the series: Corrosion and its prevention in the chemical industry. Vol. 15.)

[43] *Lehmann, H.,* and *H. Mitusch:* Feuerfester Beton aus Tonerdeschmelzzement. (Refractory concrete from smelted aluminous cement.) Goslar: Hübner 1959. (Series, Stones and Earths, Vol. 3).

[44] TGL 9271. Portlandzement. 4/1965.

[45] TGL 10573. Zemente. Physikalische Prüfung. (Cements, Physical Tests.) June, 1961.

[45a] *Bogue, R. H.:* The Chemistry of Portland Cement. 2nd Ed. New York; Reinhold Publ. Corp. 1955.

[46] *Lefol:* These. Paris 1937; quoted in *Rey, M.:* Silicates Ind. **22** (1957) p. 533.

[47] *Petzold, A.,* and *I. Göhlert:* Tonind.-Ztg. (Clay Industry News.) **86** (1962) No. 10, p. 228.

[48] *Heller, L.:* Proc. Third Symp. Chem. Cement. London 1952, p. 237.

[49] *Mčedlov-Petrosjan, O. P., V. I. Babuškin* and *G. M. Matveev:* Termodinamika silikatov (The thermodynamics of silicates.) Moscow: Gosstrojizdat 1962, p. 225, et. seq.[1]).

----

[1]) Appeared as a German translation in VEB Verlag für Bauwesen, Berlin 1965.

[50] *Šauman, Z.:* Silikaty **3** (1959) p. 46.

[51] Contribution to discussion, VII Conference of the Silicates Industry, Budapest 1963.

[52] *Duderov, G. N.:* Bull. stroit. techn. 23/24 (1946); quoted in [6].

[53] TGL 9272. Eisenportlandzement. Hochofenzement. (Iron Portland Cement. Blast Furnace Cement.) May, 1965.

[54] *Endell, K., W. Müllensiefen* and *K. Wagenmann:* Metall und Erz (Metal and Ore). **29** (1932) p. 368.

[55] *Gibbels, H.:* Feuerfester Beton auf Basis des Unterwellenborner Hochofenzements. Diplomarbeit. (Refractory concrete made with blast furnace cement from Unterwellenborn. Diploma Thesis.) Weimar 1960.

[56] *Pole, G. R.,* and *D. G. Moore:* Journ. Amer. Ceram. Soc. **29** (1946) p. 20.

[57] *Talaber, J.:* Epitöanyag **14** (1962). No. 3, p. 90.

[58] TGL 9738. Tonerdeschmelzzement. (Smelted aluminous cement.) April, 1963.

[59] *Schneider, S. J.:* Journ. Amer. Ceram. Soc. **42** (1959) No. 4, p. 184.

[60] *Kukolev, G. V.,* and *A. I. Rojzen:* Žurn. prikl. chim. **25** (1952) No. 5.

[61] *Kravčenko, I. V.:* Glinozemistyi cement (Aluminous cement). Moscow: Gosstrojizdat 1961.

[62] *Kukolev, G. V., M. T. Mel'nik* and *N. N. Šapovalova:* in: Trudy šestogo sovеščanija po eksperimental'noj i techničeskoj mineralogii i petrografii (Papers of the 6th Symposium on experimental and technical mineralogy and petrography). Moscow: Izdat. Akad. Nauk SSSR 1962, p. 228.

[63] *Arnould, J.:* Chim. et Ind. **70** (1953) No. 6, p. 1081.

[64] *Braniski, A.:* Tonind.-Ztg. (Clay Industry News). **85** (1961) No. 6, p. 125.

[65] *Budnikov, P. P.,* and *V. G. Savel'ev:* in: Trudy MChTI (Proceedings of the Mendeleev Institute). Moscow: Vol. 27 (1959), p. 272.

[66] *Nadachowski, F.:* Silicates Ind. **27** (1962) No. 7/8, p. 355.

[67] *Matveev, R. R.:* Ogneupory 1937, No. 10, p. 684; quoted in [65].

[68] *Budnikov, P. P.,* and *V. G. Savel'ev:* Ogneupory **27** (1962) No. 9, p. 412.

[69] *Savel'ev, V. G.:* Issledovanie monobarievogo aljuminata i ogneupornych betonov na ego osnove. (An investigation into monobarium aluminate and into refractory concretes made with this material.) Diss. Moscow 1962.

[70] *Butterling, B.:* Wasserglas-Säurekitte. (Waterglass acid cements.) Diss. Weimar 1962.

[71] *Dombrovskaja, N. S.,* and *M. R. Mitel'man:* Žurn. prikl. chim. **26** (1953) No. 9.

[72] *Ost-Rassow:* Lehrbuch der chemischen Technologie (Textbook of Chemical Technology). 24. Ed. Leipzig: J. A. Barth 1952, p. 489.

[73] *Ključarov, Ja. V.,* and *N. V. Mešalkina:* Cement **23** (1957) No. 5, p. 14.

[74] *Il'ina, N. V.,* and *L. I. Skoblo:* Giprocement, Trudy. (Proceedings of the Institute of Giprocement.) Vol. 26 (1963) p. 143.

[75] TGL 12375. Magnesiabinder. (Magnesia binding agents.) Sept. 1962.

[76] *Searle, A. B.:* Refractory materials: Their manufacture and uses. 3rd Ed. London: Griffin & Co. 1953, p. 731.

[77] *D'jačkov, P. N., G. G. Zagajnov, O. N. Zajkov* and *B. T. Fišel':* Ogneupory **28** (1963), No. 8, p. 361.

[78] *Greger, H. H.,* and *J. J. Reimer:* USA-Pat. 2425152 (1947).

[79] *Greger, H. H.:* Brick and Clay Record **117** (1950) No. 2, p. 63.

[80] *Kingery, W. D.:* Journ. Amer. Ceram. Soc. **33** (1950), No. 8, p. 239.

[81] *Kingery, W. D.:* Journ. Amer. Ceram. Soc. **35** (1952), No. 3, p. 61.

[82] *Gitzen, W. H., L. D. Hart* and *G. MacZura:* Amer. Ceram. Soc. Bull. **35** (1956), No. 6, p. 217.

[83] *Barta, R.:* Stavivo **39** (1961), No. 8, p. 282.

15*

[84] *Bechtel, H.*, and *G. Ploss:* Ber. dtsch. keram. Ges. (Report of the German Ceramics Society.) **37** (1960), No. 8, p. 362.

[85] *Bechtel, H.*, and *G. Ploss:* Ber. dtsch. keram. Ges. (Report of the German Ceramics Society.) **40** (1963), No. 7, p. 399.

[86] *Sheets, H. D., J. J. Bulloff* and *W. H. Duckworth:* Brick and Clay Record **133** (1958), No. 1, p. 55.

[87] *Matveev, M. A.*, and *A. I. Rabuchin:* Ogneupory **26** (1961), No. 6, p. 281.

[88] *Harris, H. M.*, and *H. J. Kelly:* Amer. Ceram. Soc. Bull. **37** (1958), No. 7, p. 307.

[89] *Dietzel, A.*, and *I. Hinz:* Ber. dtsch. keram. Ges. (Report of the German Ceramics Society.) **39** (1962), No. 12, p. 569.

[90] TGL 9397. Begriffe und Bezeichnungen der Feuerfest-Industrie. Erzeugnisarten. Febr. 1962. Hilfsstoffe. Jan. 1963. (Descriptions and Notation for the Refractory Industry. Types of products. Febr. 1962. Subsidiary materials. Jan. 1963.) TGL 99-13 Begriffe und Bezeichnungen der Feuerfest-Industrie. Roh- und Hilfsstoffe. Jan. 1963. (Descriptions and Notation for the Refractory Industry. Raw materials and subsidiary materials. Jan. 1963).

[91] *Salmanov, G. D.:* in: Issledovanija po žaroupornym betony i železobetony (Research on heat-resistant concrete and reinforced concrete). Moscow: Gosstrojizdat 1954.

[92] ...: Das Betonwerk (Concrete). **30** (1942), No. 1/2, p. 6.

[93] *Vajner, L. I.:* Prom. stroit. **31** (1953), No. 1, p. 30.

[94] *Hansen, W. C.*, and *A. F. Livovich:* Journ. Amer. Ceram. Soc. **36** (1953), No. 11, p. 356.

[95] *Celujko, E.*, and *V. S. Lavrent'e:* Stroit. mat. **2** (1956), No. 6, p. 20.

[96] *Röhrs, M.:* Baustoff-Ind. (The Building Materials Industry). **5** (1962), No. 7, p. 194.

[97] *Röhrs, M.*, and *H. Gibbels:* Baustoff-Ind. (The Building Materials Industry.) **6** (1963), No. 12, p. 376.

[98] *Mitusch, H.:* Ber. dtsch. keram. Ges. (Report of the German Ceramics Society.) **39** (1962), No. 9, p. 454.

[99] *Gibson, R. F.*, and *A. F. Old:* USA-Pat. 2903778 (1959).

[100] *Milovanov, A. F.*, and *V. M. Prjadko:* Stroit. mat. **9** (1963), No. 5, p. 28.

[101] TGL 4323. Feuerfeste Baustoffe. Schamottesteine für allgemeine Zwecke. Klassifikation. (Refractory Building Materials. Chamotte bricks for general purposes. Classification.) April 1963.

[102] *D'jačkov, P. N.:* Ogneupory **26** (1961), No. 9, p. 394.

[102a] TGL 4322. Feuerfeste Baustoffe. Silikasteine. (Refractory Building Materials. Silica bricks.) Jan. 1963.

[103] *Ludera, L.:* Zement-Kalk-Gips **12** (1959), No. 12, p. 575.

[104] *Schmid, I.:* Silikattechn. (Silicate Technology.) **12** (1961) No. 1, p. 22.

[105] *Eusner, G. R.*, and *D. H. Hubble:* Amer. Ceram. Soc. Bull. **39** (1960), No. 8, p. 395.

[106] *Kuttanova-Kresova, V.:* Stavivo **32** (1954), No. 11, p. 383.

[107] *Franke, G.*, and *F. Kanthak:* Silikattechn. (Silicate Technology.) **10** (1959), No. 9, p. 445.

[108] *Cejtlin, L. A.:* in [291], p. 103.

[109] *Siegrist, A.:* J. Constr. Suisse Romande **30** (1955) No. 12, p. 667.

[110] *Cejtlin, L. A.:* in [291], p. 98.

[111] *Epštejn, S. A.:* in [291], p. 203.

[112] *Maslennikova, M. G.:* in [291], p. 192.

[113] *Hrdlicka, L.*, and *E. Viktor:* Stavivo **35** (1957), No. 4, p. 141.

[114] *Ivanova, V. P.:* in [291], p. 127.

[115] *Petzold, A.*, and *C. Dressler:* Wiss. Zeitschr. d. Hochschule f. Architektur und Bauwesen. (Scientific Papers of the Technical University for Architecture and Building.) Weimar **10** (1963), No. 5, p. 593.

[116] *Kühl, H.:* Zementchemie. Band III. (The Chemistry of Cement, Vol. III.) 3rd Ed. Berlin: VEB Verlag für Bauwesen 1961, pp. 302, 318.

[117] *Krivickij, M. Ja.:* Žaroupornyj penobeton, ego svojstva i prigotovlenie. (Heat--resistant foamed concrete; its properties and manufacture.) Proceedings of the CNIPS, Vol. 2, Moscow: Gosstrojizdat 1950.

[118] *Martin, K.:* Versuche zur Herstellung von Feuerleichtbeton nach dem Schaumverfahren. Diplomarbeit. (Research into the manufacture of lightweight refractory concrete by the foaming process. Diploma thesis.) Freiberg/Sa. 1961.

[119] *Sadovnikov, G. A.:* Stroit. mat. 6 (1960), No. 11, p. 19.

[120] TGL 11357. Beton in aggressiven Wässern. Beurteilung des Wassers. (Concrete in aggressive water. Water assessment.) June 1962.

[121] Werbematerial der Societe Anonyme de Chaux & Ciments de Lafarge & du Teil, Marseille: Verarbeitungsanleitung Fondu Lafarge. (Publicity Brochure of the "Société Anonyme de Chaux & Ciments de Lafarge & du Teil, Marseille". Directions for Handling Lafarge Aluminous Cement.)

[122] *Koči, B.:* Inženyrski stavby 5 (1957), No. 1, p. 13.

[123] *Ludera, L.:* Feuerfeste Betone. Broschüre des Feuerfestwerkes Swidnica. (Refractory Concretes, Brochure of the Swidnica Refractory Works.)

[124] *Hummel, A.:* Das Beton-ABC. Berlin: Ernst u. Sohn 1959.

[125] TGL 11360. Sand und Kies und gebrochene Natursteine. Prüfung (Entwurf). (Sand and Gravel and Crushed Natural Rock. Testing (Design).) March, 1951.

[126] TGL 9928. Prüfung von gebrochenen Natursteinen. Beurteilung der Kornform. (Testing of Crushed Natural Rock. Assessment of Particle Shape.) Feb. 1961.

[127] TGL 0-1045. Bauwerke aus Stahlbeton. (Building structures in Reinforced Concrete.) April 1963.

[128] *Kuntzsch, E.,* and *G. Rabe:* Silikattechn. (Silicate Technology.) 10 (1959), No. 9, p. 448.

[129] *Salmang, H.:* Die Keramik. (Ceramics.) 2nd Ed. Berlin-Göttingen-Heidelberg: Springer 1951.

[130] *Bolomey, J.:* Schweiz. Bauztg. 98 (1931), No. 9, p. 106.

[131] *Bolomey, J.:* Schweiz. Bauztg. 88 (1926), No. 2, p. 56.

[132] *Mittag, C.:* Die Hartzerkleinerung. (The Crushing of Hard Materials.) Berlin-Göttingen-Heidelberg: Springer 1953.

[133] *Kirchberg, H.:* Aufbereitung bergbaulicher Rohstoffe. (The preparation of mined raw materials.) Jena: Gronau 1953.

[133a] TGL 10809. Körnungen für Sand, Kies und zerkleinerte Natursteine. (Grading of Sand, Gravel and Crushed Natural Rock.) Oct. 1962.

[134] DAMW-N 25-260. Prüfung von körnigen Stoffen. Bestimmung der Rohdichte. (The Testing of Granular Materials. Determination of Density.) Dec. 1963.

[135] *Röbert, S.:* Wiss. Zeitschr. d. Hochschule f. Bauwesen. (Technical Papers of the Technical University for Building.) Leipzig 1963, No. 12, p. 79.

[136] Feuerbeton. Gemische und Fertigbauteile — ein Fortschritt im Ofenbau. Werbeschrift der Feuerfest-Industrie der DDR. (Refractory Concrete, Mixes and Precast Units — A new Development in Furnace Construction. Publicity Brochure of the Refractory Industry of the German Democratic Republic.) 1964.

[137] *Schmeisser, E.:* Ziegelind. (Brick Industry.) 8 (1955), No. 15, p. 593.

[137a] TGL 0-1048. Bestimmungen für Betonprüfungen bei Ausführung von Bauwerken aus Stahlbeton. (Concrete testing during the construction of reinforced concrete works.) Oct. 1963.

[138] *Grün, W.:* Betonzusätze, Spezialbeton. (Concrete Additives. Special Concrete.) Düsseldorf: Verlag Baustoff-Forschung Albrecht KG. 1959.

[139] *Zahlbruckner, H.:* Radex-Rdsch. 1956, No. 6, p. 309.

[140] TGL 9233. Fertigbauteile aus Betongemischen. Abgaskanalschieberplatten für Betriebstemperaturen bis 1200 °C. (Precast Concrete Components. Drop-gates for Chimney Flues for Operating Temperatures up to 1200 °C.) June 1962.

[141] *Engelmark, C. E.:* Levindustrien **57** (1954) No. 8, p. 170.

[141a] ...: Maschinenmarkt **57** (1951) No. 28, p. 31.

[142] Pyro-Schmidt. Feuerfeste Mörtel und Stampfmassen (Werbeschrift). (Pyro-Schmidt. Refractory Mortars and Concretes (Publicity Brochure).)

[143] *Graf, O.:* Die Eigenschaften des Betons. (The Properties of Concrete.) Berlin: Springer 1950.

[144] *Reinsdorf, S.:* Leichtbeton. (Lightweight Concrete.) Vol. II. Berlin: VEB Verlag für Bauwesen 1963.

[145] *Kudrjašev, I. T.,* and *V. P. Kuprjanov:* Jačeistye betony. (Porous Concretes.) Moscow: Gosstrojizdat 1959.

[146] TGL 117-0377. Leichtzuschlagstoff-Schaumbeton. (Foamed Concrete from Lightweight Aggregates.) April 1962.

[147] *Vogrin, C. M.,* and *Heep, H.:* Paper at the Petroleum Session of the Amer. Soc. for Mech. Engineers 1955. quoted in [148].

[148] *Livovich, A. F.:* Amer. Ceram. Soc. Bull. **40** (1961) No. 9, p. 559.

[149] *Salmanov, G. D.:* in [291] p. 137.

[150] *Nisino, T.,* and *K. Moteki:* Journ. Ceram. Ass. Japan **69** (1961) No. 5, p. 130.

[151] *Alešina, O. K.:* in [291], p. 144.

[152] *Kravčenko, I. V.,* and *O. K. Alešina:* Bull. Izobret 1960. No. 15, p. 82.

[153] *Jung, V. N.,* and *A. I. Koršunova:* Stroit. mat. 1931, No. 7; quoted in [6].

[154] *Jung, V. N.:* Zement **21** (1932) p. 433.

[155] *Gitzen, W. H.,* and *L. D. Hart:* Amer. Ceram. Soc. Bull. **40** (1961) No. 8, pp. 503, 510.

[156] *King, D. F.,* and *A. L. Renkey:* USA-Pat. 2845360 (1958).

[157] *Cejtlin, L. A., A. A. Sorokin, M. F. Filičkin* and *N. F. Buntman:* Stal' 1958, No. 3.

[158] *Fedorov, A. E.:* in [291], p. 183.

[159] *Michal'čik, P. A.:* in [291], p. 192.

[160] *Wygant, J. F.,* and *W. L. Bulkley:* Amer. Ceram. Soc. Bull. **33** (1954) No. 8, p. 233.

[161] *Heindl, R. A.,* and *Z. A. Post:* Journ. Amer. Ceram. Soc. **37** (1956) No. 5, p. 205.

[162] *Heindl, R. A.,* and *Z. A. Post:* Journ. Amer. Ceram. Soc. **33** (1950) No. 7, p. 230.

[163] *Hansen, W. C.,* and *A. F. Livovich:* Amer. Ceram. Soc. Bull. **34** (1955) No. 9, p. 298.

[164] *Epštejn, S. A.:* Stroit. prom. **33** (1955) No. 7, p. 39.

[165] *Hammond, E.:* Foundry Trade J. **107** (1959) pp. 169, 203.

[166] *Gitzen, W. H., L. D. Hart* and *G. MacZura:* Journ. Amer. Ceram. Soc. **40** (1957) No. 5, p. 158.

[167] Werkstandard WST 99-12-2. Feuerfestbeton-Fertigbauteile. Technische Lieferbedingungen. (Refractory concrete precast units. Technical Specification.) Feb. 1963 (VEB Silika- und Schamottewerk Rietschen.)

[168] *Al'tšuler, B. A., G. D. Salmanov* and *A. P. Tarasova:* in: Akademija Stroitel'stva i Architektury SSSR — Naučno-issledovatel'sky institut betona i železobetona. (Reports of the Research Institute for Concrete and Reinforced Concrete.) Vol. 6 (1959) p. 136.

[169] *Schneider, S. J.,* and *L. E. Mong:* Journ. Amer. Ceram. Soc. **41** (1958) No. 1, p. 27.

[170] *Schneider, S. J.,* and *L. E. Mong:* Journ. Res. Nat. Bur. Stand. **59** (1957) No. 1, p. 1.

[171] *Guljaeva, V. F.,* and *G. D. Salmanov:* Ogneupory **28** (1963) No. 4, p. 165.

[172] Werkstandard WST 99-12-3. Feuerfestleichtbeton-Fertigbauteile. Technische Lieferbedingungen. (Precast Units of Lightweight Refractory Concrete. Technical Specification.) Feb. 1962. (VEB Silika- und Schamottewerk Rietschen.)

[173] Werkstandard WST 99-12-4. Feuerfestbeton-Fertigbauteile mit mittlerer Rohdichte. Technische Lieferbedingungen. (Precast Units of Refractory Concrete of

Medium Density. Technical Specification.) May 1963. (VEB Silika- und Schamotte-werk Rietschen.)

[174] *Nekrasov, K. D.:* in: Sbornik naučnych rabot po vjažuščim materialam. (Proceedings of the Mendeleev Institute, Moscow; Binding Agents.) Moscow: Promstrojizdat 1949, p. 83.

[175] *Haase, Th.:* Silikattechn. (Silicate Technology.) **1** (1950) No. 1, p. 5.

[176] *Rudolph, G.:* Untersuchungen zur Temperaturwechselbeständigkeit von Feuerbeton. Diplomarbeit. (Investigations into the resistance of refractory concrete to fluctuating temperatures. Diploma Thesis.) Freiberg/Sa. 1960.

[177] *Haase, Th.,* and *K. Petermann:* Silikattechn. (Silicate Technology.) **7** (1956), No. 12, p. 505.

[178] TGL 9358. Prüfung keramischer Roh- und Werkstoffe. Bestimmung der Wasseraufnahme, Rohdichte, offenen Porosität, Gesamtporosität. (The Testing of Ceramic raw materials and ceramic industrial materials. Determination of water absorption, density, open porosity and total porosity.) April 1962.

[179] *Eusner, G. R.,* and *J. T. Shapland:* Journ. Amer. Ceram. Soc. **42** (1959) No. 10, p. 459.

[180] *Paul, W. B.:* Amer. Ceram. Soc. Bull. **33** (1954) No. 4, p. 108.

[181] *Venable, C. R.:* Amer. Ceram. Soc. Bull. **38** (1959) No. 7, p. 363.

[182] *Milovanov, A. F.,* and *V. S. Zyrjanov:* Ogneupory **25** (1960), No. 5, p. 234.

[183] *Cejtlin, L. A.,* and *F. Z. Dolkart:* Ogneupory **18** (1953) No. 5.

[184] *Ludera, L.:* Zement-Kalk-Gips **14** (1961) No. 5, p. 212.

[185] *Kajbičeva, M. N., L. Ja.. Pivnik* and *N. I. Mar'evič:* Ogneupory **27** (1962) No. 4, p. 166.

[186] *Stock, D. F.,* and *J. L. Dolph:* Amer. Ceram. Soc. Bull. **38** (1959) No. 7, p. 356.

[187] *Bergman, K. G.:* in [291], p. 158.

[188] *Žicharevič, S. A., A. I. Rojzen, E. A. Ginjar, L. A. Kosyreva, A. F. Kablukovski* and *S. D. Skorochod:* Ogneupory **24** (1959) No. 7, p. 309.

[189] *Nekrasov, K. D.:* in [291], p. 107.

[190] *Gamarnik, Ja. M.:* in [291], p. 107.

[191] *Hansen, W. C.,* and *A. F. Livovich:* Amer. Ceram. Soc. Bull. **37** (1958) No. 7, p. 322.

[192] *Wygant, J. F.,* and *M. S. Crowley:* Journ. Amer. Ceram. Soc. **41** (1958) No. 5, p. 183.

[193] *Ključarov, Ja. V.:* in [291], p. 70.

[194] *Rubinštejn, V. S., I. I. Antonova* and *I. B. Veprik:* Naučnoe soobščenie VNII-Stojneft 1956; quoted in [42].

[195] *Sassa, V. S.:* in [291], p. 174.

[196] *Gajlit, A. A., N. I. Grafas, A. S. Cyganov, B. Ju. Šagalova, K. D. Nekrasov* and *V. S. Sassa:* Ogneupory **25** (1960) No. 12, p. 520.

[197] *Zalkind, I. Ja.:* in [291], p. 77.

[198] *Beljaev, A. K., N. V. Il'ina* and *L. G. Skoblo:* in [291], p. 132.

[199] *Sassa, V. S.:* Stojkost' žaroupornych betonov v rasplavach solej natrija. (The stability of heat-resistant concretes in molten sodium salts) (Scientific report of the Academy for Building and Architecture of the USSR, Vol. 3) Moscow: Gosstrojizdat 1958.

[200] *Tarasova, A. P.:* in [291], p. 151.

[201] *Il'ina, N. V.,* and *O. G. Skoblo:* Giprocement Trudy (Proceedings of the Giprocement Institute). Vol. 27 (1963) p. 107.

[202] *Maslennikova, M. G.:* Legkie žaroupornye betony na židkom stekle i portlandcemente (Lightweight refractory concretes made with waterglass and portland cement). Moscow: Gosstrojizdat 1958.

[203] *Nekrasov, K. D.,* and *M. G. Maslennikova:* Beton i železobeton **7** (1961) No. 2, p. 63.

[204] *Bron, V. A., S. P. Zamotaev, M. V. Medjakova, K. P. Semavina* and *L. B. Chorošavin:* Ogneupory **26** (1961) No. 3, p. 115.

228   References

[205] *Pirogov, A. A., Ja. R. Krass, I. E. Boriskin, D. S. Kostinskij, G. E. Socha* and *Ju. P. Evdokimov:* Ogneupory **26** (1961) No. 4, p. 176.
[206] *Pirogov, A. A.,* and *F. Z. Dolkart:* Ogneupory **25** (1960) No. 3, p. 114.
[207] *Pirogov, A. A.,* and *V. P. Rakina:* Ogneupory **24** (1959) No. 3, 125.
[208] *Pirogov, A. A.:* in [291], p. 56.
[209] *Leve, E. N.:* in [291], p. 116.
[210] *Pirogov, A. A., E. N. Leve* and *P. D. Pjatikop:* Ogneupory **25** (1960) No. 6, p. 260.
[211] *Bron, V. A.,* and *K. P. Semavina:* Ogneupory **28** (1963) No. 9, p. 385.
[212] *Cser, A.:* Epitöanyag **9** (1957) No. 6, p. 293.
[213] *Cser, A.:* Ziegelind. **12** (1959) No. 5, p. 133.
[214] *Nekrasov, K. D.,* and *G. D. Salmanov:* Ogneupory **26** (1961) No. 10, p. 417.
[215] *Amerikov, A. V.,* and *A. A. Pirogov:* Ogneupory **25** (1960) No. 11, p. 527.
[216] *Nekrasov, K. D.:* Ogneupory **27** (1962) No. 11, p. 524.
[217] *Solodennikov, L. D.:* in [291], p. 6.
[218] *Zotov, A. V.:* in [291], p. 10.
[219] *Fel'dgandler, G. G.:* Ogneupory **28** (1963) No. 7, p. 295.
[220] *Fel'dgandler, G. G.:* in [291], p. 85
[221] *Brieschke, B.:* Neue Hütte **5** (1960) No. 12, p. 734.
[222] *Ivanickij, V.:* Stroit. Gaz. **24** (1960) No. 113, p. 3.
[223] *Milovanov, A. F.,* and *V. S. Zyrjanov:* Prom. stroit. **38** (1960) No. 12, p. 32.
[224] *Nekrasov, K. D.,* and *G. D. Salmanov:* Ogneupory **25** (1960) No. 5, p. 219.
[225] *Milovanov, A. F.:* Beton i železobeton **5** (1959) No. 9, p. 417.
[226] *Zyrjanov, V. S.:* Prom. stroit. **38** (1960) No. 7, p. 45.
[227] *Issers, A. E.:* in [291], p. 46.
[228] *Besedin, S. M.:* Ogneupory **19** (1954) No. 3.
[229] *Kuntzsch, E.:* Freiberger Forschungshefte A 242. Berlin: Akademie-Verlag 1962, p. 33.
[230] ...: Ogneupory **25** (1960) No. 12, p. 531.
[231] *Murašev, V. Z., N. I. Lukaškin, K. D. Nekrasov* and *S. Krugljak:* Stroit. prom. 1949, No. 10.
[232] *Nemirovskij, Ja. M.,* and *B. A. Al'tšuler:* Stroit. prom. **31** (1953) No. 6, p. 25.
[233] *Glebov, S. V.:* Ogneupory **24** (1959) No. 6.
[234] *McCullough, D.:* Iron and Steel Engineer 1959, No. 9, p. 172.
[235] *Serebrennikov, S. S.:* in [291], p. 29.
[236] *Bel'skij, V. I.,* and *A. Kudrjacev:* Stroit. Gaz. **25** (1961) No. 89, p. 3.
[237] *Bel'skij, V. I.:* in [291], p. 34.
[238] *Kiričenko, N. D.,* and *E. V. Kočinev:* in [291], p. 213.
[239] *Hedley, C.:* Iron and Steel **33** (1960) No. 6, p. 272.
[240] *Butenko, V. A., I. E. Dudavskij, M. I. Kolesnik* and *I. N. Sokolov:* Ogneupory **28** (1963) No. 11, p. 486.
[241] *Sachlin, V. I., T. G. Šunin, A. F. Tarasov, A. M. Kulakov, K. D. Nekrasov* and *G. D. Salmanov:* Ogneupory **28** (1963) No. 8, p. 364.
[242] *Bron, V. A., L. B. Chorošavin, V. F. Isupov* and *L. Z. Kljukina:* Ogneupory **26** (1961) No. 6, p. 265.
[243] *McCullough, D.:* Proc. of Electric Furnace Steel Conf. 1956, p. 46; quoted in [220].
[244] *Al'tšuler, B. A., G. D. Salmanov, A. D. Sokol'skij* and *T. P. Karasev:* Ogneupory **22** (1957), No. 9.
[245] *Žicharevič, S. A.,* and *A. I. Rojzen:* in [291], p. 124.
[246] *Gojkolov, E. F.,* and *A. V. Zotov:* Bjull. stroit. techn. 1946, No. 12/13.
[247] *Serebrennikov, S. S.:* Cvetnye metally 1952, No. 3.
[248] *Chanin, S. E.,* and *P. A. Ivanov:* Stroit. prom. **33** (1955) No. 7, p. 38.
[249] *Sabaneev, P. F.:* Ogneupory **18** (1953) No. 10.
[250] *Šalaev, V. V.:* Metallurg 1956, No. 8, p. 28.

[251] *Šalaev, V. V., V. E. Burkser, P. P. Borodin, P. N. D'jačkov, A. K. Purgin* and *I. P. Bol'šakov:* Ogneupory **27** (1962) No. 6, p. 264.
[252] *Tolkačev, P. I.,* and *V. P. Borisov:* Stroit. prom. **35** (1957) No. 9, p. 32.
[253] *Thompson, N. M.:* Ind. Heating **26** (1959) No. 2, p. 362.
[254] *Al'tšuler, B. A.,* and *S. D. Rabinovič:* Ogneupory **24** (1959) No. 11, p. 522.
[255] *Rabinovič, S. D.:* Ogneupory **26** (1961) No. 4, p. 200.
[256] *Nekrasov, K. D.:* Stroit. prom. **37** (1959) No. 6, p. 54.
[257] *Glebov, S. V.,* and *E. A. German:* Ogneupory 1937, No. 7.
[258] *Rojzen, A. I.:* Ogneupory 1938, No. 9; quoted in [220].
[259] *Rittgen, A.:* Ber. dtsch. keram. Ges. (Report of the German Ceramics Society.) **20** (1939), p. 131.
[260] *Miehr, W.:* Sprechsaal **85** (1952) No. 12, p. 273; No. 13, p. 299.
[261] *Thompson, N. M.:* Amer. Ceram. Soc. Bull. **32** (1953) No. 1, p. 1.
[262] *Wilck, K.:* Sprechsaal **89** (1956) No. 7, p. 133.
[263] *McCullough, J. D.:* Brick and Clay Record **141** (1962) No. 6, p. 36.
[264] *Karpjuk, Ju. K.:* in [291], p. 257.
[265] *Robson, T. D.:* British Clayworker **62** (1954) p. 302.
[266] *Lobaugh, F. E.:* Brick and Clay Record, Dec. 1946; quoted in [12].
[267] *Sauzier, P.:* L'industria italiana dei Laterizi **7** (1953) No. 4, p. 173; ref. Tonind.-Ztg. **78** (1954) No. 7/8, p. 129.
[268] *Schmeisser, E.:* Ziegelind. **8** (1955) No. 24, p. 928.
[269] *Schmeisser, E.:* in: Ziegeleitechn. Jahrbuch 1957, p. 209.
[270] ...: Stroit. Gaz. **24** (1960) No. 27, p. 2.
[271] *Valeur, F.:* DRP 134820.
[272] *Beljaev, A. K.:* Cement 1949, No. 4.
[273] *Bašev, F. P.,* and *M. P. Nazarov:* Ogneupory **24** (1959) No. 11.
[274] *Beljaev, A. K.:* Cement **25** (1959) No. 4, p. 25.
[275] ...: Bau **3** (1960) No. 43, p. 11.
[276] ...: Bauinformation 1963 No. 12, p. 116.
[277] *Robson, T. D.:* Coke and Gas **14** (1952) p. 241.
[278] *Pirogov, A. A.,* and *L. A. Cejtlin:* Ogneupory 1945, No. 2/3; quoted in [220].
[279] *Burjak, P. T., S. S. Posternak* and *N. S. Gutnik:* Stroit. prom. **37** (1959) No. 12.
[280] *Posternak, S. S.:* in [291], p. 260.
[281] *Hodgson, J.:* Refractories J. **36** (1960) No. 1, pp. 14, 20.
[282] *Kuznecov, N. I.,* and *I. Ja. Zalkind:* Bjull. stroit. techn. **10** (1953) No. 13, p. 10.
[283] Rukovodjaščie ukazanija po primeneniju ogneupornych betonov v obmurovok parovych kotlov. (Guide to the use of refractory concretes in boiler linings.) Orgenergostroj 1958.
[284] *Paul, W. B.,* and *D. L. Garrison:* Radex-Rdsch. 1956, No. 2, p. 67.
[285] *Rjazanov, L. S.:* in [291], p. 295.
[286] *Vejcman, M. A.,* and *I. M. Ceperskij:* in [291], p. 208.
[287] *Petersen, P. H.:* ASTM Bull. 1955, No. 27, p. 26.
[288] *Protze, H. G.:* Journ. Amer. Concr. Inst. **28** (1957) p. 871; ref. Betonstein-Ztg. **24** (1958) No. 8, p. 336.
[289] ...: Cement, Lime, Gravel **35** (1960) No. 3, p. 70.
[290] *West, R. R.,* and *W. J. Sutton:* Amer. Ceram. Soc. Bull. **30** (1951) No. 2, p. 35.
[291] Žaroupornye beton i železobeton v stroitel'stve. (Heat-resistant concrete in building.) Moscow: Gosstrojizdat 1962.
[292] Preisanordnung Nr. 3005. Feuerfeste Rohstoffe, Erzeugnisse und Altmaterialien, Jan. 1964. Sonderdruck P 3005. Staatsverlag der Deutschen Demokratischen Republik. (Price Regulation No. 3005. Refractory raw materials, products and salvaged materials, Jan. 1964. Special publication P 3005. State publishing house of the German Democratic Republic.)

230   *References*

[293] Ukazanija po prigotovleniju i primeneniju žaroupornych betonov. (Guide to the manufacture and use of heat-resistant concretes.) Moscow: Gosstrojizdat 1958.

[294] *Valenta, D.*, and *Z. Kovarova:* Inzenyrské stavpy **13** (1965) No. 9, p. 382.

[295] *Röhrs, M., R. Müller* and *S. Plüschke:* Wiss. Zeitschr. d. Hochschule f. Architektur und Bauwesen. (Scientific papers of the Technical University for Architecture and Building.) Weimar **12** (1965) No. 1, p. 13.

[296] Neuerervorschlag Baustoffkombinat Mühlhausen und Institut für Baustoffkunde Weimar 1965. (Latest Recommendations of the Building Materials Factory, Mühlhausen and the Institute for Building Materials, Weimar 1965).

[297] *Dietrich, W.*, and *E. Kallenowsky:* Erprobung von Hochtemperaturbeton für die Steinkohlenentgasung. Bericht No. 4/1345/64 F des Instituts für Energetik Leipzig. (Tests of the use of high temperature concrete in the gasification of coal. Report No. 4/1345/64 F of the Institute for Power, Leipzig.)

[298] *Mostovoj, Ja. P., M. K. Goksadze, V. G. Sicharulidze, A. A. Čigunadze, N. G. Džinčaradze* and *B. V. Garišvili:* Ogneupory **29** (1964) No. 10, p. 471.

[299] Wissenschaftlich-technischer Bericht des Forschungsinstituts für Stahlbeton. (Economic and technical report of the Research Institute for Reinforced Concrete.) Moscow: No. 607—62; quoted in [6].

[300] Instrukcija po proektirovaniju̇ i ustrojstvu protivokorrozijnoj saščity vytjažnych trub predprijatij s agressivnymi sredami. (Instructions for the design and construction of corrosion protection linings for extract ducts in contact with aggressive media.) SN 163-61. Moscow: Gosstrojizdat 1961.

[301] *Gburek, A.*, and *R. Röhlig:* Baustoff-Ind. **9** (1966) No. 1, p. 23.

[302] *Milovanov, A. F.:* Žarostojkij železobeton. (Heat-resistant reinforced concrete.) Moscow: Gosstrojizdat 1963.

[303] *Milovanov, A. F.*, and *V. M. Prjadko:* Beton i železobeton **9** (1963) No. 5, p. 215.

[304] *Moskvin, V. M.*, and *A. P. Podval'nyj:* Beton i železobeton **7** (1961) No. 6, p. 246.

[305] *Nekrasov, K. D.*, and *A. P. Tarasova:* in [306], p 28.

[306] Stroitel'stvo promyšlennych pečej i dymovych trub. (The construction of industrial furnaces and chimneys.) Vol. 6. Central'noe bjuro techničeskoj informacii. Moscow 1964.

[307] Primenenie žarostojkogo betona v stroitel'stve pečej. (The use of heat-resistant concrete in furnace construction.) Central'noe bjuro techničeskoj informacii. Moscow 1964.

[308] *Haubenreisser:* Vortrag anlässlich eines Erfahrungsaustausches mit *K. D. Nekrasov.* (Paper on an exchange of information and experience with *K. D. Nekrasov.*) Meissen 1964.

[309] *Preisser:* Vortrag anlässlich eines Erfahrungsaustausches mit *K. D. Nekrasov.* (Paper on an exchange of information and experience with *K. D. Nekrasov.*) Meissen 1964.

[310] *Majzel', I. L.*, and *M. F. Sucharev:* Žaroupornyj teploizolacionnyi perlitobeton. (Heat-resistant thermally insulating pearlite concrete.) Moscow: Strojizdat 1965.

[311] *Vasil'ev, E. S.:* in [307], p. 155.

[312] *Vorslov, V. P.:* in [306], p. 48.

[313] *Agurin, A. P.:* in [306], p. 55.

[314] *Trinker, B. D.:* in Sbornik techničeskoj informacii. (Collected technical information.) Series V: Special'nye raboty v promyšlennom stroitel'stve. (Special works of industrial construction.) Vol. 4 (8). Moscow 1961, p. 50.

[315] *Pomorceva, E. N., V. A. Medvedev* and *S. R. Zamjatin:* Ogneupory **29** (1964) No. 7, p. 308.

[316] *Brodulin, A. I.*, and *M. G. Levin:* Stal' 1961, No. 1; quoted in [315].

[317] *Barta, R.*, and *B. Mokry:* Hutnik **5** (1955) No. 12, p. 360.

[318] *Krickij, G. G.:* in [306], p. 77.

[319] *Kolb, L.:* Silikattechn. (Silicate Technology). **16** (1965) No. 5, p. 160.

[320] *Protze, H. G.:* Cf. [288]; ref. in Bauplanung-Bautechnik **11** (1957) No. 10, p. 453.

[321] *Cook, M. D., C. P. Cook* and *D. F. King:* Amer. Ceram. Soc. Bull. **42** (1963) No. 9, p. 486; No. 11, p. 694; **43** (1964) No. 5, p. 380.

[322] *Wygant, J. F.:* Amer. Ceram. Soc. Bull. **42** (1963) No. 5, p. 317.

[323] *Livovich, A. F.:* Amer. Ceram. Soc. Bull. **45** (1966) No. 1, p. 11.

[324] *Bordzenko, L. S.:* Sbornye bytovye peči i dymovye truby. (Built-up domestic stoves and chimneys.) Moscow: Gosstrojizdat 1962.

[325] *Kuntzsch, E.:* Neue Hütte **10** (1965) No. 5, p. 275.

9. Index